Zebra Stripes

Zebra Stripes

TIM CARO

The University of Chicago Press   Chicago and London

The University of Chicago Press, Chicago 60637
The University of Chicago Press, Ltd., London
© 2016 by The University of Chicago
All rights reserved. Published 2016.
Printed in the United States of America

25 24 23 22 21 20 19 18 17 16   1 2 3 4 5

ISBN-13: 978-0-226-41101-9 (cloth)
ISBN-13: 978-0-226-41115-6 (ebook)
DOI: 10.7208/chicago/9780226411156.001.0001

Library of Congress Cataloging-in-Publication Data

Names: Caro, T. M. (Timothy M.), author.
Title: Zebra stripes / Tim Caro.
Description: Chicago: The University of Chicago Press, 2016. | Includes
    bibliographical references and index.
Identifiers: LCCN 2016019386 | ISBN 9780226411019 (cloth: alk. paper) |
    ISBN 9780226411156 (e-book)
Subjects: LCSH: Zebras–Color. | Stripes. | Protective coloration (Biology)
Classification: LCC QL737.U62 C38 2016 | DDC 599.665/71472–dc23 LC
    record available at https://lccn.loc.gov/2016019386

♾ This paper meets the requirements of ANSI/NISO Z39.48-1992
    (Permanence of Paper).

How often have I said to you that when you have eliminated the impossible, whatever remains, however improbable, must be the truth?

—SHERLOCK HOLMES IN *THE SIGN OF FOUR* BY SIR ARTHUR CONAN DOYLE, 1890

Contents

Color plates follow page 110.

Figure fm.1 Plains zebra showing an area of the body with thin white and black stripes. The face is often close to the ground while the animal is feeding. (Drawing by Sheila Girling.)

Preface and acknowledgments

We often learn about the external appearances of animals as children but are rarely told the reasons that animals wear their characteristic colors and patterns. Most mammals have drab brown or gray coats, but a few of them, especially the primates and some squirrels and marsupials, sport brightly colored hair or exhibit conspicuous sexual skin. Other groups, like skunks and porcupines, are often white and black and advertise their defenses, but these are exceptional (Caro 2013). Zebras have a unique striped black and white coat and are not obviously defended; thus, the evolution of zebra stripes poses intriguing questions for biologists and the public alike. If we could discover why zebras have black and white stripes, we could solve a long-standing and fascinating evolutionary mystery and would certainly enhance delight in the natural world by linking science with the beauty of these animals. Fortunately our scientific understanding of the evolution of animal coloration is growing very rapidly at present—I personally have been thinking about mammalian coloration for 20 years—and there are now at least four groups of scientists working on the problem of zebra coloration, so the solution may be in our grasp.

I started to work on the conundrum of why zebras have stripes in August 2005, taking Alfred Russel Wallace's (1896) advice about the importance of looking at utility of coloration in the field, advice that he offered in many of his writings. He even wrote about this issue specifically in discussing zebra stripes: "And we may also learn how impossible it is for us to decide on the inutility of any kind of coloration without a careful study of the habits of the species in its native country" (p. 220). I studied this problem in zebras' natural habitat, almost exclusively in Katavi National Park in western Tanzania, where between 14,000 and 29,000 plains zebras live (Caro 2008). Every year in each dry season and in some wet seasons, I would test one or more hypotheses for the function of striping but generally returned to my home university in the United States with negative results. This was frustrating, and only when we used a comparative phylogenetic analysis of coat coloration across equid species and subspecies did we make a breakthrough. In one sense, then, I need never have left the University of California library on the Davis campus for Africa, but my testing of every hypothesis for striping using observational or experimental techniques in the field does serve two purposes. First, it brings the debate out of the armchair or laboratory into the field, and this is the first strength of this book. Second, it tests every hypothesis systematically using quantitative data. This is important for moving a field forward, in this case beyond more implausible ideas for why zebras are striped to asking precise questions. The comprehensive scope of the book is its second strength. The third strength is that I bring a wide range of methods and approaches to bear on a problem that does not lend itself to easy experimentation.

I need to thank many people for help along the way. First, I want to thank my mother, Sheila Girling, for the beautiful drawings at the beginning of each chapter; this is the third book of mine she has illustrated, and I am most grateful. Unfortunately, she passed away before publication. Rickesh Patel drew Figure 8.1. I thank the University of California and National Geographic Society for funding.

For permission to work in Tanzania, I am grateful to the Tanzania Wildlife Research Institute (TAWIRI), Commission for Science and Technology (COSTECH), and Tanzania National Parks (TANAPA), and in particular I would like to thank Katavi chief park wardens, the late Ally Kyambile, Stefano Qolli, Chris Timbuka, and John Gara for being so welcoming to their park. For absolutely essential logistical support, I particularly want to thank the late Oska Ulaya, as well as Amisa Msago at the research base in Kibaoni, and car mechanics Linus Mwitwa and Geofrey Milambo in Sumbawanga, and Japhert Gama in Mpanda.

At the Tierpark Zoo in Berlin, I thank Bernhard Blaszkiewitz (former director) and Florian Sicks (curator of mammals) as well as the Wissenschaftskolleg zu Berlin. I thank the Chicago Field Museum, Los Angeles Natural History Museum, California Academy of Sciences, and Museum of Vertebrate Zoology at Berkeley for access to pelts and skulls and Hannah

Walker for thoughtful work in these museums. For extensive library and web-based data collection in the United States, I am extremely grateful to Hannah Walker and also to Amanda Izzo, and Zoe Rossman.

For help in the field, I thank Monique Borgerfoff Mulder, Andrew Bower, Elizabeth Carabine, Barnabas Caro, Caroline Chumo, Peter Genda, Jacob Mwalyolo, Damien Overton, Jon Salerno, and Laura Young; for advice about equipment, I thank Henrik Akerblom, Christian Kiffner, and Gail Patricelli; for advice about field techniques, I thank Ken Britten, Muffadel, and Craig Packer (and Craig for lending me his IR camera); for insect identification, I am grateful to Steve Mihok and Danielle Whisson. For analyses in Chapter 2, I thank Amanda Melin, Don Kline, and Chihiro Hiramatsu; for those in Chapter 7, I thank Ken Britten and Timothy Thatcher; and for those in Chapter 8, I thank Amanda Izzo, Bobby Reiner, and Ted Stankowich; for help with historical references, I acknowledge Tom Schoener. For general discussions about why zebras have stripes, I am grateful to Ken Britten, Monique Borgerhoff Mulder, my late father, Sir Anthony Caro, Amanda Izzo, Brenda Larison, Craig Packer, Dan Rubenstein, Tom Sherratt, and Martin Stevens. Finally, I thank Sue Cohan, Mary Corrado, Christie Henry, Logan Ryan Smith and Gina Wadas at the University of Chicago Press, and Innes Cuthill, Dan Rubenstein, Martin Stevens, and an anonymous reviewer for constructive comments on the book manuscript.

Figure 1.0. A quagga, a subspecies of plains zebra now extinct. The striped forequarters but uniform hindquarters attracted much attention in western Europe in the 1800s, and now in South Africa, where efforts are under way to breed zebras to mimic quagga coloration. (Drawing by Sheila Girling.)

ONE

Stripes and equids

1.1 The question of stripes

Zebras are one of the most visually arresting animals in nature because they have contrasting and regular black and white striped coats. They look like no other mammal and have a strikingly different colored coat from the familiar but closely related domestic horse. To most people, they are exceptionally beautiful. As far back as 1824, Burchell, after whom the plains zebra is sometimes named, wrote, "I stopped to examine these zebras with my pocket telescope: they were the most beautifully marked animals I had ever seen: their clean sleek limbs glittered in the sun, and the brightness and regularity of their striped coat, presented a picture of extraordinary beauty, in which probably they are not surpassed by any quadruped with which we are at present acquainted. It is, indeed, equaled in this particular, by the dauw (mountain zebra) whose stripes are more defined and regular, but which do not offer to the eye so lively a colouring" (p. 315).

Outside of Africa, zebras have long been popular curiosities. For instance, Grevy's zebras were brought to ancient Rome to draw chariots in circuses in AD 211–217, and quaggas were used to pull carriages in Hyde Park, London, in the 1800s (MacClintock 1976). Yet despite the animals' attraction, the riddle of why zebras have stripes has never been satisfactorily solved (Cloudsley-Thompson 1999; Ruxton 2002; Caro 2011). Long debated as far back as the nineteenth century by Wallace (1867a, 1879) and Darwin (1871), it has fomented much discussion by other great biologists (Poulton 1890; Beddard 1892; G. Thayer 1909; Mottram 1916; Cott 1940) and keen observers of natural history (Kipling 1902; Selous 1908; Cloudsley-Thompson 1984; Kingdon 1979; Morris 1990). Now, there are many intriguing ideas, but few have been tested experimentally, and it is difficult to take insights from similarly colored species because repeatedly alternating black and white stripes are found in so few vertebrates. Examples include the zebra duiker (see Appendix 1 for scientific names); okapi; some snakes, such as the desert banded snake and California king snake; and fishes, including the zebra shark, zebra moray, and zebra red dorsal. Unfortunately, the adaptive significance of striping patterns in snakes and fishes is also poorly understood (e.g., Seehausen, Mayhew, and Van Alphen 1999; Allen et al. 2013; Kelley, Fitzpatrick, and Merilaita 2013). In short, the reasons that three species of equid have striped pelage have not lent themselves to easy investigation. This book tries to fill that gap.

In this chapter, I first outline the functional hypotheses and associated mechanisms that have been proposed for why zebras have evolved black and white striped pelage. This task has been conducted by others (e.g., Cloudsley-Thompson 1984; Kingdon 1984; Morris 1990; Ruxton 2002; Egri, Blaho, Kriska, et al. 2012), but my purpose here is to make this list comprehensive and recast these ideas into larger functional categories that then enable us to use ecological data to bear on the evolutionary drivers of striping. I then briefly describe the natural history and external coloration of zebras and other equids since many interspecific comparisons will be made in this book, and finally summarize what is known about zebra hair.

1.2 Hypotheses for striping in equids

1.2.a Antipredator hypotheses

By far the most renowned and popular ideas as to why zebras are striped center on stripes reducing the likelihood of zebras being hunted or killed by predators. These come in several guises.

Crypsis

The first notion as to why zebras have stripes is for concealment against predators. Close up and in daylight, zebras' contrasting black and white livery is very conspicuous to the human eye, but at a distance and under low illumination, when predators hunt, zebras might be difficult to see, and early ideas about zebra coloration focused on this. Wallace (1896), the father of the field of animal coloration, remarked, "It may be thought that such extremely conspicuous markings as those of the zebra would be a great danger in a country abounding with lions, leopards and other beasts of prey; but it is not so. Zebras usually go in bands, and are so swift and wary that they are in little danger during the day. It is in the evening, or on moonlight nights, when they go to drink, that they are chiefly exposed to attack; and Mr Francis Galton, who has studied these animals in their native haunts, assures me, that in twilight they are not at all conspicuous, the stripes of white and black so merging together into a gray tint that it is very difficult to see them at a little distance" (p. 220). We now call this type of coloration background matching. G. Thayer (1909) argued that stripes conceal zebras in "reeds and grasses, or even bare-limbed bushes and low trees, or sand streaked with shadows of any of these plants, or quiet water striped with their reflections—its obliterative effect must be almost perfect" (p. 138). It is not clear whether Thayer's purported mechanism was background matching or disruptive coloration where false edges break up the outline of the animal, because he stated, "The stripes . . . still play their true obliterative part 'cutting the beast to pieces'" (p. 138). Certainly Cott (1940), in his benchmark and influential book on animal coloration, as well as subsequent authors (e.g., Matthews 1971; McLeod 1987), thought that stripes broke up the continuous surface contour of the animal and thereby masked the margin of the body or the body's appendages; these biologists also subscribed to the idea that zebras were difficult to see at a distance (see also Hingston 1933). Cotton (1998) was of the opinion that stripes might produce a shimmering effect in a heat haze. In addition, Cott and others such as Cloudsley-Thompson (1980) proposed that narrowing of the dark flank stripes and lighter belly acted as countershading, making the body appear flat and difficult to recognize as a prey object (see also Mottram 1916). Thus, there are actually three different mechanisms that have been proposed for striping being a form of crypsis: background matching, disruptive coloration, and countershading.

Aposematism

Darwin remarked, "The zebra is conspicuously striped, and stripes on the open plains of South Africa cannot afford any protection" (Darwin 1906, p. 832). Darwin was likely referring to protective coloration, used in Victorian times in the sense of background matching. Conspicuous coloration, however, can also offer protection in another way by advertising antipredator defenses as Wallace (1896) and Poulton (1890) recognized, and this theme was picked up by other authors. For example, shortly afterward, Selous (1908, p. 22), the infamous big game hunter, wrote, "Never in my life have I seen the sun shining on zebras in such a way as to cause them to become invisible or even in any way inconspicuous on an open plain, and I have seen thousands upon thousands of Burchell's zebras." Similarly, Roosevelt (1910), another hunter and Selous's companion in the bush, described zebras coming down to a waterhole to drink: "They were always very conspicuous, and it was quite impossible for any watcher to fail to make them out" (p. 561), and "Never in any case did I see a zebra come down to drink under conditions which would have rendered it possible for the most dull-sighted beast to avoid seeing it" (p. 561).

In mammals, warning coloration, also called aposematism, usually takes the form of black and white coloration, most memorably in skunks and stink badgers, where it is associated with production of toxic secretions, but it may also advertise other defenses such as quills (Caro 2013). Simply by analogy, then, the black and white external appearance of zebras might advertise dangerous kicks and bites that equids are known to deliver (Matthews 1971; Kruuk 1972) and thereby warn predators not to attack them.

Confusion

Another constellation of ideas for stripes thwarting predation is through confusing the predator. Several mechanisms have been proposed, any of which might be involved. Possibly stripes might make it difficult to count zebras accurately if the stripes on adjacent individuals appear to form a continuous line, making it difficult to distinguish individuals from one another. Another possibility is that if members of a herd flee together, the stripes on different individuals appear to merge into a single line of black and white stripes. This might make it challenging for a predator to isolate an individual visually and so target its final attack accurately (Kruuk 1972; Eltringham 1979; Morris 1990). Alternatively, stripes might make it difficult to follow a single zebra in a herd that is fleeing erratically. Another form of confusion is that white and black stripes might shimmer and apparently shift, making it difficult for a predator to concentrate on either stationary or fast-moving prey (Morris 1990). Yet another possibility is that stripes cause motion dazzle, defined as "markings that make estimates of speed and trajectory difficult by the receiver" (Stevens and Merilaita 2011, p. 5). Here stripes, vertical, oblique, and horizontal, are believed to interfere with the observer's perception of the speed at which a zebra moves (Morris 1990). Experimental work with objects moving across a computer screen demonstrates that humans perceive zigzag striped patterns as moving more slowly than plain objects (Scott-Samuel et al. 2011; Stevens, Yule, and Ruxton 2008; Stevens et al. 2011; Hughes, Troscianko, and Stevens 2014; but see von Helversen, Schooler, and Czienskowski 2013). Optical illusions could even be involved (see Hall et al. 2016), including the "wagon-wheel effect," where vertical flank stripes generate motion signals in the opposite direction to actual movement, or the "barber's pole illusion," where diagonal stripes on the rump produce motion signals 50°–60° away from the actual movement direction (How and Zanker 2013). A final idea is that stripes running perpendicular to the body's outline cause it to appear larger than similar-sized objects that have stripes running parallel to the body's outline (Cott 1940; Vaughan 1986; Morris 1990) or that vertical stripes cause the object to appear shorter and fatter (Helmholtz [1867] 1962). Either effect might confuse the predator and cause it to misjudge the position of the area of the body on which it plans to land, thereby preventing it from making proper contact with prey.

Quality assessment

Patterns of striping are different in every individual zebra and offer the opportunity for humans to keep track of an individual in a herd. If predators can do the same and are loath to pursue individuals in good physical condition, then it might be beneficial for individuals in good condition to wear stripes to make sure they can be personally recognized. Under this hypothesis, it is conjectured that stripes are a signal that advertises an individual's quality (Ljetoff et al. 2007), a type of antipredator defense that in other mammals is usually mediated through behavior (Caro 2005).

1.2.b Antiparasite hypotheses

The second major but less well known hypothesis comes from experiments showing that biting flies are less likely to land on black and white striped surfaces than on uniform surfaces or objects of other colors (R. Harris 1930). Females of several biting insect taxa require a blood meal to reproduce and may cause great annoyance while feeding on hosts, including equids. A handful of experimental studies have investigated this. Waage (1981) found that black and white striped boards caught fewer *Glossina morsitans* and *G. pallidipes* tsetse flies (glossinids) in the field than all-black boards or all-white boards whether the boards were moving or stationary; and similar results were found for tabanids (horseflies and deerflies). He argued that stripes might obliterate the stimulus of the body edge or reduce the amount of contrast of the animal's form against its background, or else be too narrow to elicit attraction. He believed that attraction from a distance rather than the landing response was influenced by striping. Brady and Shereni (1988) found similar results for *Glossina morsitans* and *Stomoxys calcitrans* (stable flies), demonstrating a reduction in landings as stripe number and stripe thinness increased. G. Gibson (1992) showed that tsetse flies were much less attracted to vertically striped and gray targets than black or white surfaces and that horizontally striped targets caught <10% as many flies as any other target. She suggested that horizontal stripes might appear to a tsetse fly to be inconspicuous patches of dark and light, while some edges of the animal perpendicular to the alignment of stripes might be difficult to detect due to absence of lateral inhibition. It may therefore be noteworthy that stripes on all parts of a zebra's body except the forehead lie perpendicular to the outline of the animal's body. Egri, Blaho, Kriska, and colleagues (2012) showed that tabanids are less likely to land on black and white striped surfaces than uniform black or (arguably) white surfaces, that this effect is more marked as stripe width declines, and that the most effective deterrent stripe widths used in their experiments matched the range of stripe widths found on the three species of zebra. They also demonstrated that tabanids are attracted to horizontally polarized light and argued that, due to backscattering, white hair should reflect light with a lower degree of polarization than black or brown hair (Horvath et al. 2008; Horvath et al. 2010). Biting flies transmit several dangerous diseases—most famously tsetse flies transmit trypanosomiasis (sleeping sickness), known to be lethal to domestic horses (Molyneux and Ashford 1983)—so stripes might reduce the likelihood of zebras catching diseases. In summary, there are at least two related ideas here: that stripes impede the landing responses of glossinids, tabanids, or *Stomoxys* biting flies and that they achieve this by altering the brightness or polarization properties of reflected light.

1.2.c Communication hypotheses

The third major hypothesis for the evolution of striping in equids is that stripes act as markers in the context of social interactions. The first idea is that stripes serve to distinguish zebras from other species. Wallace (1891) noted that "the stripes therefore may be of use by enabling stragglers to distinguish their fellows at a distance." (p. 368), and Darwin (1906), quoting Hunter, remarked that "a female zebra would not admit the addresses of a male ass until he was painted so as to resemble a zebra" (p. 825). Also striping patterns might help zebra species to distinguish one another in areas of sympatry—namely, between Grevy's and plains zebra in northern Kenya, and between plains and mountain zebra in southern Africa—because hybridization can occur. Certainly different patterns of striping on the rump in each species are quite obvious (Morris 1990).

A second idea is that stripes serve to direct allogrooming behavior (mutual grooming) of conspecifics toward the subject's mane, neck, and withers, areas that cannot be reached by the subject itself (Kingdon 1984). Stripes are conjectured to mimic folded skin that appears when the neck of a uniformly colored herbivore is twisted or bent, so stripes might guide a conspecific to these particular areas of the body that are in need of ectoparasite removal.

A closely related idea concerns social cohesion (Kingdon 1979). Allogrooming might promote social bonding as suggested for domestic horses (Kimura 1998) and help keep subgroups together when they aggregate in large herds characteristic of some zebra species (Ruxton 2002). As an offer of support, Kingdon cites observations that zebras come to a halt parallel to each other and in close proximity to each other more than do domestic horses, and that captive Grevy's zebras with their thin stripes prefer to stand next to a panel of fine rather than thick stripes. Stripes, he argued, would be unable to operate in this way in equid species with shaggy winter coats (Kingdon 1984; but see Morris 1990).

Third, striping is believed to foster communication between conspecifics. First, unique striping patterns present on each individual might aid in individual recognition (Darwin 1871; Klingel 1977; Morris 1990). More specifically, patterns on the rump might help other zebras follow one another (Kingdon 1984). Alternatively or additionally, stripes might allow individuals to be clearly seen at a specific distance and so enable them to keep apart (Cloudsley-Thompson 1999).

A fourth idea postulates that some aspect of striping, perhaps stripe thicknesses or brightness of pelage, advertises the condition of the individual that might be used in contest competition or in mate choice.

1.2.d Thermoregulation hypothesis

The final major hypothesis states that white and black stripes could act as a cooling device. Black stripes absorb radiation, whereas white stripes reflect it (Baldwin 1971), and it has been proposed that alternating black and white stripes set up convection currents over the surface of the animal and thereby cool it (Morris 1990). Surface temperature measurements affirm that black stripes heat up more quickly and obtain higher equilibrium temperatures than white stripes (Louw 1993). Moreover, black stripes have more subcutaneous fat beneath them, preventing heat penetrating too deeply (Cotton 1998). Certainly, the greater proportion of black stripes on the flank and the greater proportion of white hair on the rump on plains and mountain zebras (but not on Grevy's zebras) would allow individuals behavioral options for warming in the morning by turning their lateral surface to the sun and for cooling in the afternoon by directing their rump to the sun (Louw 1993). Black and white pelage must reflect heat in different ways, but the questions of whether these have an overall cooling effect and by what mechanism and whether they actually reduce core temperature are open.

1.3 Equid evolution

The Equidae are one of three families in the Perissodactylae (odd-toed ungulates), along with the Rhinocerotidae and the Tapiridae. Modern equids are medium- to large-sized herbivores, 250–400 kg, reaching over a meter at shoulder height (Table 1.1). They have long necks and heads, incisor teeth for clipping and molars for grinding grasses, and their middle digits bear the weight of each hoofed foot (Rubenstein 2006). The evolutionary history of the equids is well documented (Lydekker 1912; Simpson 1951; Bennett 1980) yet still an active field of inquiry (e.g., Weinstock et al. 2005; Steiner and Ryder 2011; Oakenfull and Clegg 1998; Rubenstein 2011). For example, over evolutionary time we know that equids show a progressive trend toward larger body size and grazing (Stirton 1940; Webb 1977). We recognize that early evolution took place in North America. We know that equids are descended from the diminutive *Protorohippus* (*Hyracotherium*) present in the Eocene, 55 MYA, which resembled a modern small carnivore but that had lost two side toes from the hindfoot and one from the forefoot. Subsequently, the slightly larger *Orohippus* appeared in the fossil record in the middle Eocene. By the Oligocene, 34 MYA, *Mesohippus* and *Miohippus* had appeared, with larger and more complex teeth. Then, during the Miocene, 24 MYA, the grasses (*Graminae*) evolved and equids radiated enormously, evolving continuously, growing tall high-crowned molariform teeth and long strong incisors to crop grass. Such taxa included *Hypohippus* and *Parahippus* as well as *Anchitherium*, which had lost the fourth toe on the front foot and had less independent metacarpals and metatarsals, traits indicative of a running lifestyle. *Anchitherium* migrated from North America to Europe 20 MYA. Later in the Miocene, around 17 MYA, *Merychippus* also evolved in North America and gave rise to extant equids. In the Pliocene, 4.5 MYA, *Pliohippus* appeared in North America, followed by the genus *Equus*, which evolved into 25–30 species,

Table 1.1. Equid measurements.

	Head/body length (cm)	Tail length (cm)	Tail/body ratio* (%)	Shoulder height (cm)	Weight (kg)	Reference
Plains zebra		47–56			175–385	[a]
	217–246					[b]
				130	290–340	[c]
				127–140	175–322	[d]
					220–250	[e]
Mean†	**232**	**52**	**22.2**	**130**	**270**	
Mountain zebra	210–260	40–55			240–372	[b]
		40–47		116–128.5	204–260	[f]
		40		130	350–430	[c]
				146	230–386	[d]
	215	50			260	[g]
Mean	**225**	**45**	**20.1**	**133**	**299**	
Grevy's zebra	250–300	38–60		140–160	352–450	[b]
	250–275	50–75		125–150	352.9–430.9	[h]
					386–430	[e]
Mean	**269**	**63**	**23.3**	**144**	**400**	
African wild ass	195–200	40–45		115–125	270–280	[i]
					250	[b]
				110–120		[c]
Mean	**200**	**43**	**21.3**	**118**	**263**	
Asiatic wild ass	200–250	30–49		100–142	200–260	[b]
					290	[g]
	200–220	43–48		126–130		[i]
Mean	**218**	**43**	**19.5**	**125**	**260**	
Kiang	118–214	32–45		132–142	250–400	[i]
Mean	**166**	**39**	**23.2**	**137**	**325**	
Przewalski's horse	220–280	92–111		120–146	200–300	[b]
	210	90				[g]
	180–280				200–350	[i]
Mean	**230**	**96**	**41.7**	**133**	**263**	

Note: For sources that included both male and female measurements, the ranges of the measurements were combined into one range.
*Tail/body ratio was calculated by dividing the tail length by the body length, and multiplying by 100.
†Means were calculated by averaging the midpoints of the ranges.
[a]Sundaresan, Fischhoff, Dushoff, and Rubenstein (2007).
[b]Nowak (1999).
[c]Stuart and Stuart (1997).
[d]Kingdon (1997).
[e]Estes (1993).
[f]Penzhorn (2013).
[g]Macdonald (1984).
[h]S. Williams (2013).
[i]Moehlman, Kebede, and Yohannes (2013).
[j]A. Smith and Xie (2013).

some of which crossed into Eurasia and later Africa. By 3½ MYA, the *Hippotigris* subgenus had appeared and spread to Eurasia and thence to Africa. Present-day horses, zebras, and asses are all thought to be derived from *Hippotigris*, but extinct species belonging to these groups have been unearthed in Europe that are 1½ to 2 million years old, suggesting a less linear evolution but more explosive diversification with rapid extinctions (Rubenstein 2011). Although the crucible of evolution of equid ancestors was in North America, equids disappeared from there by the end of the last ice age and were found only in the Old World until very recently; zebras are only found in Africa.

Nowadays there are only 7 extant equid species: the plains zebra (*Equus burchelli* or *E. quagga* [which I refer to as the former in this book]); the mountain zebra (*Equus zebra*); Grevy's zebra (*Equus grevyi*); the African wild ass (*Equus africanus*); the Asiatic wild ass, or onager (*Equus hemionus*); the kiang, or Tibetan wild ass (*Equus kiang*); and Przewalski's horse (*Equus ferus przewalskii*) (Groves 2013) (Plate 1.1). The relationship between equid clades (branches in the tree of life) is still under discussion, for it hinges on the type and region of DNA used in different studies (Rubenstein 2011), but classically it has been assumed that zebras and then the asses diverged from the ancestral true horses, with onagers separating from African wild asses half a million years ago (Groves 2013; Prothero and Schoch 2003). Rapid speciation across the clade may have been facilitated by swift changes in chromosome numbers that differ between species (Bush et al. 1977; Wichman et al. 1991; Trifonov et al. 2008).

These 7 species are made up of 23 subspecies when the recently extinct quagga *E. b. quagga*, Burchell's zebra *E. b. burchellii*, and Atlas wild ass *E. a. atlanticus* are included (Appendix 1 in Moehlman 2002a), although there is debate about subspecific differences (Groves and Bell 2004; Kruger et al. 2005; Orlando et al. 2009; Orlando et al. 2013). All modern equids are grazers (Schulz and Kaiser 2013) that live in open semiarid areas (Moehlman 1985) (Table 1.2) and spend much of the day feeding. They have long skulls with powerful jaws, and their hind-gut digestive system, consisting of a simple single chambered stomach with a long cecum and large intestine, allows them to digest large quantities of coarse vegetation of poor nutritional value. All equids are polyestrus, meaning they go into heat several times each year, and they exhibit breeding seasons in temperate regions. Gestation lasts 11–13 months, depending on the species (Asa 2002). Their breeding systems fall into two categories (Klingel 1975): adult females in plains zebras, mountain zebras, Przewalski's horses, and feral domestic horses form stable relationships with other mares and a stallion or two that leads to a stable breeding and social group. This is the

Table 1.2. Distribution and habitats of equids.

	Distribution	Habitat
Plains zebra	Kenya, Tanzania[a] South of Sahara[c] Southeastern Sudan to southern Africa and Angola[e]	Grasslands, steppes, savannah[b] Open woodlands[d] Bushland[f]
Mountain zebra	South Africa[g]	Arid mountains, bushy karoo shrubland[b] Barren, rocky uplands, arid plains[h] Rugged, mountainous areas[d]
Grevy's zebra	Northern Kenya[i] Southern Somaliland, southern Ethiopia[c] Laikipia-Samburu ecosystem of central Kenya[j]	Bush/grass mosaics, grass on deep sand, hardpans, sumplands, seasonally waterlogged plains[b] Thornbush country, barren plains, foothills[h] Semiarid desert plains, dry/open woodland, savannah[d]
African wild ass	Northeastern Sudan, northeastern Ethiopia, western Djibouti[k] Danakil Desert, northern Eritrea, northern Somalia[l]	Semidesert grasslands, dwarf shrublands typified by aloes and euphorbias, rocky hills[b] Arid, rugged hill country and desert plains[d]
Asiatic wild ass	Syria to Mongolia[c] Karakum desert in Turkmenia; areas of Iran, India, Russia, Tibet[l]	High upland, rocky outcrops, mountain massifs[m] Drought deserts, semidrought deserts, steppes, mountain deserts[n]
Kiang	China, India, Nepal, Pakistan[o] Ladak, Tibet[p] Tibetan plateau in China[q]	Alpine meadows, alpine steppes, open terrain, plains, basins, broad valleys, hills, grass and sedge[h] Elevated plateaus[p]
Przewalski's horse	Mongolia[r]	Semidesert, steppe flatland[s] Saline high steppes, sand dunes, rocky dry lands[t] Deserts, hilly lands, grassland[n]

[a]Hack, East, and Rubenstein (2002).
[b]Kingdon (1997).
[c]Matthews (1971).
[d]Stuart and Stuart (1997).
[e]Estes (1991).
[f]Klingel (2013).
[g]Novellie et al. (2002).
[h]Estes (1993).
[i]S. Williams (2002).
[j]Sundaresan, Fischhoff, Hartung, et al. (2007).

[k]Moehlman (2002b).
[l]Klingel (1977).
[m]Reading et al. (2001).
[n]Helin, Ohtaishi, and Houji (1999).
[o]Shah (2002).
[p]Lydekker (1912).
[q]Xu et al. (2013).
[r]Wakefield et al. (2002).
[s]Bökönyi (1974).
[t]Mohr (1971).

so-called (in regard to equids) Type I sort of social organization: harem defense polygyny in which males control access to females (Linklater 2000). In Type II resource defense polygyny, males control access to females indirectly by monopolizing critical resources. In these equids, female Grevy's zebra and female asses do not form stable long-term relationships with other females or stallions except during parturition, early maternal care, or mating (Klingel 1977; Rubenstein 1986). Some Type II populations occasionally exhibit Type I social organization (Linklater 2000). Equids are monomorphic; in other words the sexes are approximately the same body size. Zebras live 20–25 years in the wild (Duncan and Groves 2013). I now discuss the behavioral attributes and appearances of each species in turn, because these prove relevant for several of the functional hypotheses for zebra striping that follow later in the book.

1.3.a Plains zebra

The plains zebra (carrying 44 chromosomes [Wichman et al. 1991]) consists of five subspecies that live in southern and eastern Africa (Table 1.2) and two extinct subspecies that lived in southern Africa (Hack, East, and Rubenstein 2002). One of these was the extinct quagga that is now regarded as a subspecies of plains zebra (Higuchi et al. 1984; Lowenstein and Ryder 1985; Leonard et al. 2005). Plains zebras occupy areas of coastal plains up to 4000 m (Stuart and Stuart 1997). Body size increases from north to south across the continent. Currently, between 700,000 and 1 million plains zebras are found in the wild (Klingel 2013). The species inhabits grasslands and grassland-bushland mosaics and dry *Acacia* and *Brachystegia* woodland; individuals have to drink each day, so they do not move far from water. Overlapping home ranges can reach 2000 km² (Klingel 1969). The species is highly social, with a single male and a stable harem of 1 to 6 unrelated adult females and their offspring (Klingel 1969); young males and males without harems live together in loose aggregations of up to 15 individuals (Table 1.3). Both of these sorts of groups come together in large groups of hundreds of individuals when grazing or moving to new areas, and in these circumstances harem holders may form coalitions to drive bachelors away from their females. Offspring are born throughout the year and are able to follow their mothers quickly after birth. Both males and females disperse from harems at puberty. Grooming of other individuals is reportedly relatively common (Table 6.1 in Chapter 6).

In the northern part of the species' range, the Grant's subspecies (*E. b. boehmi*) occupies parts of Tanzania, Kenya, Sudan, Ethiopia, Uganda, Somalia, Burundi, and Rwanda. The Upper Zambezi subspecies (*E. b. zambeziensis*) is found in Zambia, Democratic Republic of Congo, and Angola. Crawshay's subspecies (*E. b. crawshayi*) occupies Zambia, Malawi, and Mozambique, and Chapman's subspecies (*E. b. chapmani*) lives in Mozambique, Zimbabwe, and Botswana. All subspecies are heavily striped, with stripes on the neck and flank lying vertically and those on the rump lying

Table 1.3. Group sizes in equids and their social organization.

	Mean group	Max group	Aggregation
Type I			
Plains zebra	6.0[a]	10.8[b]	Hundreds[c]
Mountain zebra	4.9[d]	12.5[e]	35[f]
Przewalski's horse	10.5[g]	23.0[h]	30[i]
Type II			
Grevy's zebra	6.3[j]		175[k]
African wild ass	5.7[l]		50[m]
Asiatic wild ass	13.3[n]		412[o]
Kiang	12.0[p]		465[q]

[a]Mean 8.2, 5, 5.1, 4.4, 4.6, 4.7, 4.5, 4.2 (Hack, East, and Rubenstein 2002); mean 7.7 (Klingel 2013); mean 4–6 (Stuart and Stuart 1997); mean 1–6, max bachelor 15 (Groves 1974); mean 6 (Kingdon 1997); mean 2–6, bachelor 2–15 (Estes 1993); mean 15, mean bachelor 2–4, max bachelor 15 (Matthews 1971); mean 5–6, mean bachelor 3–6, max bachelor 15 (Estes 1991); mean 2–6, max 16 (Linklater 2000); mean 5.6 (Grubb 1981).

[b]Max 14, 10, 11, 10, 8, 9, 8, 11 (Hack, East, and Rubenstein 2002); max 16 (Linklater 2000).

[c]Aggregation hundreds (Klingel 1969); aggregation hundreds (Schaller 1972).

[d]Mean 5 (Kingdon 1997); mean 4.7 (Estes 1993).

[e]Max 12 (Kingdon 1997); max 13 (Skinner and Chimimba 2005).

[f]Aggregation 40 (Stuart and Stuart 1997); aggregation 30 (Kingdon 1997).

[g]Mean 6–7 (Groves 1974); mean 6–15 (Helin, Ohtaishi, and Houji 1999); mean 8–20 (Linklater 2000); mean 11.1 (Kaczensky et al. 2008).

[h]Max 23 (Kaczensky et al. 2008).

[i]Aggregation 10–50 (Mohr 1971).

[j]Mean 4–11 (Estes 1993); mean 5.1 (Sundaresan, Fischhoff, Dushoff, and Rubenstein 2007).

[k]Aggregation 150 (S. Williams 2013); aggregation 200 (Kingdon 1979).

[l]Mean 5 (Moehlman, Kebede, and Yohannes 2013); mean 6.3 (Klingel 1977).

[m]Aggregation 50 (Stuart and Stuart 1997); aggregation 49 (Klingel 1977).

[n]Mean 5–15 (Helin, Ohtaishi, and Houji 1999); mean 4–35, 4–7, 3–18 (Reading et al. 2001); mean 16.3 (Klingel 1977); mean 2–39, 1–34 (Bouskila et al. 2013); mean 6.3 (Feh et al. 2001).

[o]Aggregation 200–300 (Helin, Ohtaishi, and Houji 1999); aggregation 135 (Klingel 1977); aggregation 850 (Feh et al. 2001).

[p]Mean 2–15 (Schaller 1998); mean 5–10 (Groves 1974); mean 8.9 (Estes 1993); mean 3–43 (Xu et al. 2013).

[q]Aggregation 300–400 (Shah 2002); aggregation 261, 500–1000 (Schaller 1998); aggregation 500 (Estes 1993).

horizontally (Plate 1.1*A*). In the more southerly parts of the continent, however, subspecies have stripes that are more yellowy than white and show shadow stripes between the black stripes, especially on the rump. Moreover, they have fewer or no stripes on their legs and belly (Roosevelt and Heller 1914). These are the Damara or Wahlberg's subspecies (*E. b. antiquorum*), living in Botswana, Namibia, Swaziland, Lesotho, and South Africa, and the extinct Burchell's (*E. b. burchellii*) and quagga (*E. b. quagga*) subspecies from South Africa. Brown pelage but no stripes was found on the rumps of these last two subspecies (Kingdon 1979; Groves 2002).

1.3.b Mountain zebra

The mountain zebra (chromosome number 32) has a thick upright mane, tufted tail, and dewlap. There are two subspecies: Hartmann's zebra (*E. z. hartmannae*) lives in western Namibia, extreme southwest Angola, and northwestern South Africa and currently numbers 25,000 individuals or more (Table 1.2). The Cape mountain zebra (*E. z. zebra*) was found principally in the western and eastern Cape Provinces of South Africa (Novellie et al. 2002) and currently totals just 1500 individuals in 20 widely scattered reserves (Penzhorn 2013). The mountain zebra is classified as Endangered by the International Union for the Conservation of Nature. The species inhabits rugged areas up to 2000 m in the mountainous regions of southwest Africa, and some populations move between plains and desert and high coastal dunes that can be 120 km apart. Mountain zebras selectively graze coarse material and drink daily. The species shows harem defense, one stallion with four or five mares (Table 1.3) that usually occupy home ranges of 9.4 km^2 on average (Penzhorn 2013). Males without a harem form unstable bachelor groups; their aggregations sometimes reach 30 individuals but are usually smaller (Table 1.3). Mutual grooming is common (Table 6.1 in Chapter 6).

Both subspecies are striped all over the body except for the belly, which is white; stripes on the rump are very broad (Plate 1.1*B*). Body stripes connect to a dorsal stripe in a gridiron fashion in some individuals (Penzhorn 1988). Cape mountain zebras have wider black or brown stripes than Hartmann's, and the mane extends less far forward between the ears.

1.3.c Grevy's zebra

Grevy's zebras (chromosome number 46) formerly ranged from Eritrea, eastern Ethiopia, and western Somalia to central Kenya but are now found only in limited areas of central and northern Kenya, with an isolated population remaining in Ethiopia (Table 1.2). Grevy's zebra is the largest zebra (Table 1.1), with broad ears and a high, prominent mane. Males are 10% heavier than females. Approximately 2700 only remain in the wild (S. Williams 2013), and the species is classified as Endangered. Grevy's zebras graze and occasionally browse in semiarid and arid shrubland or grassland (Sundaresan, Fischhoff, Hartung, et al. 2007), but the rate at which they need to drink varies between one and five days depending on whether they are lactating. Dominant breeding males defend 2–12 km^2 territories year-round through which females pass to reach water, or defend grassy areas with light tree cover, which are attractive to nonlactating females (S. Williams 2002; Rubenstein 2006; Rubenstein 2010). Females give birth throughout the year, are solitary or live in very small groups with attendant offspring, but may also form large groups of 175 or so with other females and with whom they are in reproductive synchrony (Sundaresan, Fischhoff, Dushoff, et al. 2007) (Table 1.3). They may move over 10,000 km² in search of food and water, and crèches of offspring are sometimes left alone when mothers leave for waterholes. Mutual grooming is rare (Table 6.1 in Chapter 6).

Grevy's zebras have thin stripes all over the body, thinner than those on the flank and rump of the plains and mountain zebra; and their lateral stripes taper ventrally, leaving the belly white. On the hindquarters the flank stripes curve forward, the hindleg stripes arch upward, and a third series of concentric arches spreads forward and downward from the tail (Plate 1.1C). The muzzle is brown with a white margin.

1.3.d African wild ass

The African wild ass (chromosome number 62–64, owing to polymorphisms) is a small equid with a tufted tail and consists of two extant subspecies: the Nubian wild ass (*E. a. africanus*), living in northeast Sudan, and the Somali wild ass (*E. a. somaliensis*) in Eritrea, northeastern Ethiopia and the Ogaden, western Djibouti, and northern Somalia (Moehlman 2002c) (Table 1.2). A third subspecies, the Atlas wild ass (*E. a. atlanticus*), inhabited northeast Algeria, Morocco, and Tunisia until AD 300. The African wild ass currently lives at low densities, due to either paucity of food resources or persecution, and is classified as Critically Endangered, with fewer than 2000 left in the wild. The species inhabits rocky desert environments but requires access to surface water every three days and is found normally within 30 km of it (Moehlman, Kebede, and Yohannes 2013). Females usually live in small groups of six (certainly fewer than 10) individuals that include young, other females, and sometimes males; aggregations may reach 50 individuals, however (Table 1.3). Males set up resource territories close to water through which females must pass in order to drink, and they defend females from other males when females are in estrus (Klingel 1977; Moehlman 1998; Moehlman, Kebede, and Yohannes 2013). Grooming of other individuals is uncommon (Table 6.1 in Chapter 6).

Nubian wild asses have a gray-reddish coat, whereas Somali wild asses are buff or yellow in the summer, but both turn grayish in the winter. Both

15

subspecies have a lighter belly, often with a sharp boundary between a darker dorsum (back) and lighter ventrum (belly). The legs are lighter too but in the Somali subspecies have thin black or dark brown stripes, most prominent below the carpus and tarsus ("knees") (Plate 1.1D). The muzzle is light colored; there is a black mane, black dorsal stripe, and short hairy black brush to the tail. The Nubian wild ass has a black transverse stripe of variable length across the shoulders. According to Roman mosaics, such as that in Bona, Algeria, the Atlas subspecies had leg stripes and a shoulder stripe (Hemmer 1990).

1.3.e Asiatic wild ass

The Asiatic wild ass (chromosome number 54–56) once ranged into western Europe in the late Pleistocene but during the thirteenth century retreated to the Middle East, ranging from Arabia through Turkmenistan and as far east as Mongolia. Isolated populations of the khur (*E. h. khur*) occur in northwest India, the onager (*E. h. onager*) in Iran, the kulan or khulan (*E. h. kulan*) in Turkmenistan, and the dziggetai (*E. h. hemionus*) and Gobi khulan (*E. h. luteus*) in Mongolia (Table 1.2). The smaller Syrian wild ass (*E. h. hemippus*) became extinct in 1927. Poaching and competition with livestock threaten many of these remaining populations, and the khur and onager are Endangered. Many reintroductions have been attempted in Iran, Turkmenistan, and Kazakhstan (Groves 1974; Feh et al. 2002), and the total wild population is around 8000. Asiatic wild asses live in highland and lowland deserts (Table 1.3), and although they are predominantly grazers, they may browse succulents during the dry season; they can regularly go without water or can drink water with high salinity (Bahloul et al. 2001). Breeding is seasonal from April through September, with the rut occurring in early summer. Females group into small bands of 2–5 individuals or more and have nonexclusive home ranges of 5000 km^2 or more (Kaczensky et al. 2008). Males either show territorial resource defense encountering females that move into their territories, or else they encounter female groups as they roam (Schaller 1998). Nonetheless, Type I behavior of a stallion with small groups of mares has also been observed (Feh, Boldsukh, and Tourenq 1994; Zhirnov and Ilyinsky 1986). Aggregations of many hundreds of individuals may form temporarily (Feh et al. 2001), and mutual grooming is common (Tables 1.3 and 6.1).

There is considerable individual variation in color of the dorsum, ranging from pale gray through to brown to reddish, depending on subspecies and season (Plate 1.1E). Rump, ventrum, legs, chest, and muzzle are lighter in color, with a sharp line between dorsum and ventrum in some individuals. There is a broad dark dorsal stripe, with a white border in some subspecies. The pelt undergoes a molt lasting 40–45 days in spring, resulting in a 1.8–2.4 cm long coat, but in winter the thick yellow-brown coat is almost twice this length.

1.3.f Kiang

The kiang, or Tibetan wild ass (chromosome number 50–52), has a big head and mane; it is the largest species of wild ass and is slightly sexually dimorphic in size (Table 1.1). Three subspecies are recognized (but see Schaller 1998): the eastern (*E. k. holderi*), the western (*E. k. kiang*), and the southern (*E. k. polydon*) kiang. The species' range includes parts of India, Pakistan, Nepal, possibly Bhutan, but principally China, with all subspecies being found on the Tibetan plateau (Table 1.2). The total population size is relatively large, over 50,000 individuals (Shah 2002; Schaller 1998), but competition with livestock is a threat. Kiang feed on sedges and grasses in open grassy terrain, but their habitat extends to desert steppe and to high altitudes up to 4000 m (Table 1.3). They have keen senses, smelling a person at 400 m and seeing one at 1.5 km; their only predators are wolves. The social organization is similar to Grevy's zebra and the two ass species, with males holding isolated territories through which females pass, although stallions are observed with mares in small herds (Schaller 1998). Large aggregations of several hundred females and their attendant young congregate on good pasture in winter and autumn (Schaller 1998), whereas males live alone or in small groups of up to 10 animals or so. The rut occurs in August and September. Mutual grooming is uncommon in the wild (Table 6.1 in Chapter 6).

The coat of the kiang is rich chestnut, turning from darker brown in winter to reddish brown in summer following molt. The winter coat begins to grow in August, with hairs reaching 3.5–4.5 cm long; an 80-day molt starts in April and finishes in July, resulting in short 1.4–1.6 cm long hair for only a month or so (Groves 1974). The end of the muzzle, insides of the ears, rings around the eyes, legs, belly, chest, and ventral portion of the nape are all white, and a sharp dividing line separates the color of ventral pelage from that of the dorsum. A dark dorsal stripe extends from the mane to the tail (Plate 1.1*F*).

1.3.g Przewalski's horse

Przewalski's horse, or takhi (chromosome number 66), may have originally ranged across the steppes of central Asia, China, and western Europe.

In the last two centuries, they have been confined to the high steppe of southwest Mongolia, the Dzungarian Gobi, where they became extinct in the wild in 1969 (Table 1.2). Recently, they have been reintroduced into several sites in Hungary, Mongolia, China, Kazakhstan, and Ukraine (Wakefield et al. 2002; Robert et al. 2005; Tatin et al. 2009). The species inhabits open plains and semiarid areas; it needs to drink daily (Scheibe et al. 1998). They exhibit a Type I harem defense mating system, with a stallion controlling around five or six mares or more that give birth in late spring. Mutual grooming occupies as much as 1% of the day in captive animals (Table 6.1 in Chapter 6).

Przewalski's horse has dun or sandy-bay pelage, becoming yellowish on the belly (Plate 1.1G). The tops of the legs are light grayish but quickly darken lower down. In some individuals, between three and ten dark stripes are present on the carpus or more usually on the tarsus. The species has a dark brown erect mane and no forelock. The upper part of the tail has short guard hairs, and the distal portion has long, flowing hair; a dark dorsal stripe runs from the mane along the back to the tail tuft. Przewalski's horses shed a 2.5–7.0 cm long thick winter coat, tail hair, and mane in early spring; long hair regrows in September.

1.3.h Other equids

Several subspecies of *E. ferus* have recently become extinct. The steppe tarpan (*E. f. ferus*) was a small mouse-colored horse with long, thick hair with an ash-gray belly and black shanks; it was first recorded in 1769 in Eurasia (Groves 1974). It became extinct in the Caucasus around 1850, with the last one dying in Ukraine in 1918, although contemporary Polish working ponies look very similar (Prothero and Schoch 2003). Forest tarpans or koniks, *E. f. sylvaticus*, were smaller than steppe tarpans and assumed a white coat in winter. This subspecies was successfully bred in captivity and has been reintroduced in Europe. Other subspecies of *E. ferus* may have lived recently in Alaska and Siberia (*E. f. alaskae*) (Groves 1974).

Horses (*Equus caballus*; chromosome number 64) were domesticated around 5000 years BP. They vary greatly in size depending on the race but have longer manes than any of their nondomesticated counterparts, and a greater proportion of the tail has flowing hair. Wild horses roam open and mountainous temperate grasslands in many parts of the world (Linklater 2000), notably mustangs in North America (Berger 1986), brumbies in Australia, feral horses in Britain, and Camargue and Haflingers in the Camargue and Tyrol, respectively. Domestication probably started in Ukraine and Kazakhstan (Olsen 2006; Warmuth et al. 2012) and was followed by a rapid increase in diversity of pelage colors (Ludwig et al. 2009); the genetics of domestic horse pelage are well understood (e.g., Sponenberg 2009; Pruvost et al. 2011). Donkeys (*Equus asinus*; chromosome number 62) are derived from the Nubian wild ass (Kimura et al. 2010), and feral asses (escaped donkeys) are found in Australia, whereas burros are seen in Mexico and the United States and elsewhere.

In sum, three of the seven extant species are heavily striped, and there is subspecific variation in striping. Striping is absent or greatly reduced in the other four equid species. Why? The question has perplexed biologists for years, with many reasons being put forward, and the rest of this book examines these systematically. But before closing, I digress briefly to discuss the biology of zebra hair.

1.4. Zebra hair

Striping patterns in plains zebras are heritable (Parsons, Aldous-Mycock, and Perrin 2007), and it is easy to see that mothers and offspring have similar coat patterns, although no formal work has been carried out. During embryogenesis, unpigmented precursors of pigment cells called melanoblasts migrate from the neural crest of the embryo to the epidermis and into developing hair follicles. Early on in fetal development of the zebra at 3–5 weeks, groups of about 20 cells (0.5–1.0 cm wide), spaced roughly evenly at about 0.4 mm apart, start to produce melanosomes, so presumably local cellular differences early on in development are responsible for the characteristic asymmetrical striping patterns seen in adults (Bard 1981). Subsequently, mature pigment cells known as follicular melanocytes synthesize melanin pigment and package it into melanosomes, which they move to cortical and medullary keratinocytes for deposition into developing hair shafts. Melanocytes can produce eumelanin (black or brown) as found in zebras or phaeomelanin (red or yellow), but some melanocytes contain fewer or smaller melanosomes, which may be defective. If this occurs, there is reduced transfer of melanin into the keratinocytes, and this results in the production of gray or white hair. Accurate migration of melanoblasts and functioning of melanocytes therefore determine external coloration patterns (Slominski et al. 2004; Yamaguchi, Brenner, and Hearing 2007; Mills and Patterson 2009).

More generally, the pattern of stripe development in zebras and other species can be explained through a theoretical reaction-diffusion model where a two-dimensional array of cells producing activation and inhibition are in diffusion contact with each other and are constrained by boundary conditions (Turing 1952). These kinetics presumably control melanocyte differentiation in cells, but cellular dynamics (the relationship between follicular melanocytes, matrix keratinocytes, and dermal papilla fibroblasts) are complex. Stripes of different width and orientation form as different parts of the fetus grow at different rates, even though black and white hairs per se do not grow until six months into development (Bard 1977).

Hair fibers themselves are made up of an external cuticle and an inner cortex. The cuticle is composed of flat cells that overlap one another like roof shingles, and these make the fiber appear as a nested set of cones. The central cortex forms the bulk of the fiber and has a pigmented core, the medulla, which determines hair color. Any light entering the hair is refracted and diffused and is then reflected off the rear surface of the hair shaft back into the cortex again, possibly becoming chromatic or colored (Lechocinski and Breugnot 2011). These processes give zebra pelage its color, black in those hairs where melanin absorbs all wavelengths, white in hairs that lack melanin where few wavelengths are absorbed, or brown in hairs where eumelanin or phaeomelanin absorbs other wavelengths. Brown hairs are found in young animals and are patchily distributed within some white stripes.

Three observations point to zebras having dark coats with white stripes: white stripes and white bellies develop subsequent to dark pigmentation in the embryo (Prothero and Schoch 2003); rare abnormally colored black zebras with white spots are occasionally seen; and the quagga that lacked distal stripes had a dark unpatterned rump (Rau 1974).

1.5 Conclusion

Equid coloration falls into two sharply defined groups: the three striped zebra species and the rest (see Rubenstein 2011 for superb photographs of equids). The task of identifying the function of striping in zebras is difficult, however, not only because hypotheses are so wide ranging, including antipredator and antiparasite defenses, communication, and physiological mechanisms, but also because only two have received more than passing attention. Humans find moving striped objects difficult to target accurately on a computer screen, suggesting a possible motion dazzle effect, and glossinid tsetse flies, *Stomoxys* stable flies, and tabanid horse- and deerflies, all of which bite, are less likely to land on black and white striped than on uniform surfaces. Moreover, there are only seven species from which to make comparisons, and wild equids do not lend themselves to easy experimentation!

Furthermore, there are other questions that have garnered less attention, including why the three species have different striping patterns (Ruxton 2002), why different subspecies have different aspects of their body striped, and why sympatric artiodactyls (even-toed ungulates) do not have similarly striped coats. Fanciful ideas about zebras' ancestors being too fat and splitting their coats (Hadithi and Kennaway 1984), falling into a fire after a fight (Greaves 1988; Stewart 2004; Mostert 2012), painting themselves to avoid heat and lions, being painted through bars, wearing pajamas or their own shredded shadows (Fontes and Fontes 2002), being beaten by a monkey (Sanchez 2011), exchanging coats for horns with an oryx (Riggs 2014), or becoming striped after "standing half in the shade and half out of it, and what with the slippery-slidy shadows of the trees falling on them" (Kipling 1902, p. 34) are clearly incorrect, but scientists cannot yet claim that they have much more convincing explanations for why zebras are striped. The reason that zebras are striped is still enigmatic and captivating and constitutes one of the great mysteries of biology.

Figure 2.0. Two plains zebras among trees. Early discussions of striping centered on whether stripes afforded zebras crypsis in this environment. (Drawing by Sheila Girling.)

TWO

Predation and crypsis

2.1 Background matching

The most long-standing hypothesis for zebra coloration is that stripes are a form of camouflage. This is rather a surprising argument because everyone agrees that zebras are highly conspicuous close-up, yet here is a suggestion that zebras are not conspicuous or are even cryptic in certain habitats, at a distance, or under poor lighting conditions (Mottram 1916). First, biologists have long remarked on the resemblance between the repeated pattern of stripes on zebras and the vegetation of the habitats in which they live (Wallace 1891; G. Thayer 1909; Cloudsley-Thompson 1969; Lythgoe 1979; Morris 1990). In grassland, the chief similarity is between the vertical stripes on the lateral surface (flank) of the zebra and the vertical orientation of tall grass stems or perhaps their vertical shadows in the early morning and late evening. Nonetheless, it is recognized that the widths of stripes are larger than those of grass stems and the horizontal stripes on the rump and legs do not match the vertical shadows of grasses. In woodland, the similarity appears superficially stronger, as black vertical and horizontal stripes might resemble dark vertical tree trunks of certain widths and their low horizontal branches; white zebra stripes might appear similar to shafts of bright light between trunks and branches. In indirect support of this idea, the great hunter and naturalist Selous (1908) remarked on the absence of stripes on the hindquarters of quaggas that live in relatively treeless areas.

Two other counterintuitive ideas are that the black and white stripes fade into a gray hue either under poor lighting conditions or else at a distance (Galton 1851), making it difficult to distinguish the animal from the background haze. For example, Cott (1940, p. 94) wrote, "Under certain conditions, that is to say, in full sunlight and in open country, the Zebra may be a conspicuous enough object. But in the dusk, when he is liable to be attacked, and in country affording thin cover, he is one of the least easily recognized game animals."

Background matching is one of several forms of crypsis (Endler 1978) and is defined as "where the appearance generally matches the colour, lightness and pattern of one (specialist) or several (compromise) background types" (Stevens and Merilaita 2009a). Thus, we might expect zebras to resemble their background perfectly or, as a compromise solution, to resemble several backgrounds imperfectly (Merilaita, Lyytinen, and Mappes 2001; Merilaita and Lind 2005; Houston, Stevens, and Cuthill 2007). Therefore, we would suppose zebras to live in environments with prominent black and white linear backgrounds or else to occupy several backgrounds somewhat similar to their striping pattern.

2.1.a Initial discomfort with the idea

Three sets of observations and a mathematical model undermine the argument for this type of crypsis. First, background matching is usually enhanced through immobility (Poulton 1890; Cott 1940) because movement is detected more easily by predators (Ioannou and Krause 2009). And as Schaller (1972, p. 255) remarked, "A herd of zebra is not difficult to see, especially when animals are moving, which is usually the case." By extension, background matching is most effective if individuals live alone or in small groups, where the probability of any group member moving is reduced. Moreover, solitary animals are less likely to be seen than groups because they take up less of a predator's visual field. Indeed, cryptic markings and being solitary are often associated in nature (Cott 1940; Edmunds 1974; Ruxton, Sherratt, and Speed 2004). Therefore, if striping is a form of background matching, we would expect zebras to be solitary, or live in small groups of two or three, yet they live in larger groups, with the exception of male Grevy's zebras (Table 1.2).

Second, the behavior of zebras does not enhance crypsis because zebras do not become immobile in the face of danger. Consider immature age-sex classes of birds and mammals that are particularly vulnerable to predation. They are often cryptically colored and remain completely immobile under dangerous circumstances; examples include altricial nestlings temporarily left alone in nests on the ground, or "hider" ungulate calves that lie out in the absence of their mother (Gochfeld 1981; Stoner, Caro, and

Graham 2003). Similarly, adults of many cryptic herbivore species, such as cervids with spotted pelage and bovids with spotted or striped coats, freeze on detecting a predator (Caro et al. 2004; Broom and Ruxton 2005; Eilam 2005). Thus, if zebra stripes were a means to match the background, we might expect zebra mothers to leave their neonates hidden and adult zebras to remain immobile upon detecting predators in their vicinity. This is not what is observed, however. Zebra foals follow their mothers almost immediately after birth; zebra herd size actually increases in the presence of large carnivores (Creel, Schuette, and Christianson 2014); and adult zebras run off a short distance and turn to face lions and spotted hyenas, or even approach them (Kruuk 1972), rather than freezing or hiding by crouching on the ground. Roosevelt (1910, p. 561) commented, "The zebra never skulks, and, like most of the plains game, it never, at least when adult, seeks to escape observation—indeed in the case of the zebra (unlike what is true of the antelope) I am not sure that even the young seek to escape observation."

Third, zebra coloration is not appropriately cryptic for the habitats in which they live. All equids are grazers inhabiting desert, semidesert, steppe, open plains, rocky, or lightly wooded habitats (Table 1.2), so we might expect them all to be of a light uniform hue to match their background just like artiodactyls that live in open habitats and deserts that have pale, white, creamy, yellow, or light gray coats (Stoner, Caro, and Graham 2003; Allen et al. 2012). Yet only the asses sport such coats. Zebras have a different livery despite living sympatrically (in the same area) with bovids and being preyed upon by some of the same carnivores.

Last, Fourier analysis, the study of the way general functions may be approximated by sums of simpler trigonometric functions, indicates that zebra stripes show spatial frequencies that are unlikely to be present against their natural backdrop. This makes the animal stand out from the background especially at higher spatial frequencies (Godfrey, Lythgoe, and Rumball 1987). Nonetheless, despite these four misgivings, the crypsis idea needs examination.

2.1.b Detecting zebras

Sightings during the day

If stripes are a form of background matching, zebras should be more difficult to see in nature than sympatric nonstriped herbivores of similar size. I thought I should examine the ease with which plains zebras and other species can be seen throughout the day but pay particular attention to evenings, nights, and early mornings, the times that large carnivores hunt (Hayward and Slotow 2009). First, I compared sighting distances of 16 species of herbivores that I recorded during vehicle transects in Katavi National Park, western Tanzania, a protected area that contains both miombo woodland and open floodplains (Caro 1999a). This was part of an extensive conservation research program in which I tried to assess how different forms of protection affect large mammal populations around Tanzania's third largest national park (Caro 1999c). I drove six transects at <10 km/hr along minor tracks in the park once per month at a minimum of 18 days apart (to ensure independence) for 14 months over an 18-month period in 1995–1996. Transects were started between 06.30 and 07.00 hrs and halted around 10.30 a.m. During each transect I recorded the number of individuals of each species of large and midsized mammal that I saw, up to 500 m from either side of the transect (Caro 1999b). At each sighting I noted the number of animals in the group (defined as <50 m apart) and estimated their perpendicular distance from the transect; estimated distances were subsequently checked and revised using markers set out at known distances at camp (Scott, Ramsey, and Kepler 1981). Sightings of zebras were subsequently separated according to the seven different vegetation types as classified by Pratt, Greenway, and Gwynne (1966) and Kikula (1980).

Data showed that the majority (77.6%, N = 214 sightings) of plains zebra sightings were made in grassland (Table 2.1), although this did not differ from expectations based on habitat availability [1]. In Katavi, zebras generally congregate on open floodplains during dry season months [2]. In intermediate periods, between dry and wet seasons, they occupy both woodland and plains habitats, while in the wet season they are more likely to be found in woodlands than expected [2]. Similarly, in Serengeti Na-

Table 2.1. Observed number of sightings of plains zebras separated by vegetation type in Katavi National Park (N = 214 sightings).

	W1	W2	W3	B	BG	WG	G
Number of sightings	0	23	21	4	13	15	138
Percentage sightings	0	10.7	9.8	1.9	6.1	7.0	64.5
Percentage vegetation type	0.4	7.0	15.0	1.1	3.1	9.0	64.5

Source: From Caro (1999a).
Note: Column headings are as follows: W (woodland): trees up to 18 m, canopy cover >20%, crowns may touch, grasses and herbs present; W1: tree canopy >70%; W2: tree canopy 50%–69%; W3: tree canopy 20%–49%; B (bushland): densely growing woody vegetation of shrubby habit, low stature <6m in height, canopy >20%; BG (bushed grassland): grassland with bushes and shrubs of canopy cover 2%–20%; WG (wooded grassland): grassland with tree cover 2%–20%; G (grassland): dominated by grass, tree canopy <2% (after Pratt, Greenway, and Gwynne 1966; Kikula 1980).

tional Park, plains zebras also occupy both sorts of habitats: the long and short grass plains for 7 months of the year and the woodlands for 5 months (Maddock 1979). Elsewhere they occupy open scrub and avoid riverine habitats (Thaker et al. 2010), and they move on to open plains when lions are in the vicinity (Valeix et al. 2009). As its English name implies, the plains zebra spends a good proportion of the year in open habitat, and together with its large body size, this makes opportunities for crypsis difficult (Stankowich and Caro 2009).

Distances at which groups and individuals were sighted along these transects were estimated to the nearest 10 m up to 500 m away (Tasker et al. 1984). Zebra sightings were principally made close to the track, but 19.7% were made at >250 m away (Figure 2.1). Across species, average sighting distances were primarily driven by species' group sizes rather than body size [3], suggesting that grouping was a better determinant of sighting probability, at least for human observers. I therefore compared the distances at which zebras were sighted to other herbivore species observed in roughly similar-sized groups (zebra [striped] $\bar{X} = 10.9$ individuals, impala [brown] $\bar{X} = 10.1$, reedbuck [brown] $\bar{X} = 6.9$, waterbuck [gray] $\bar{X} = 6.1$, and elephant [gray] $\bar{X} = 6.1$) (Figure 2.1), although group cohesiveness may

Table 2.2. Global detection probabilities (scaled for swamp, grassland and woodland), truncation distances in meters, and number of encounters with the truncation distance.

	Detection probability P_a	P_a 95% CI	Truncation distance	Number of encounters
Topi	0.28	0.18–0.44	450	29
Giraffe	0.37	0.30–0.46	400	59
Warthog	0.21	0.16–0.28	360	67
Zebra	0.38	0.29–0.50	350	49
Impala	0.35	0.26–0.46	340	42
Puku	0.52	0.30–0.91	300	11
Elephant	0.48	0.36–0.64	270	31
Buffalo	0.52	0.37–0.73	230	23
Roan antelope	1.0	1.00–1.00	210	6
Waterbuck	0.55	0.33–0.92	160	14
Hartebeest	0.55	0.33–0.94	140	12
Reedbuck	0.35	0.23–0.52	140	26
Greater kudu	0.43	0.19–0.99	120	6
Sable antelope	0.66	0.36–1.00	115	12
Eland	1.0	1.00–1.00	100	8
Duiker	0.36	0.30–0.43	100	73
Small antelope[a]	0.51	0.38–0.68	68	34
Bushpig	0.85	0.43–1.00	64	12
Bushbuck	0.61	0.47–0.80	60	26

Source: From Waltert et al. (2008).

[a]Not identified to species level; potentially includes bush duiker, Sharpe's grysbok, steinbok, and oribi.

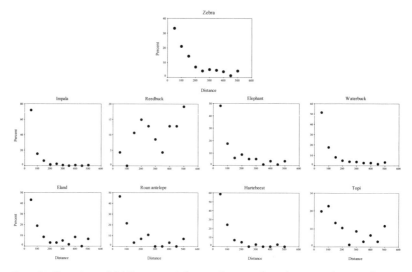

Figure 2.1. Percentage of sightings seen at distances in meters from the center of transect lines driven throughout the Katavi ecosystem summing monthly transects over a 14-month period (from Caro 1999b). Zebras (N = 228 sightings of individuals or groups) are shown above plotted against distances in 50 m intervals or less. Below are four species with similar group sizes to zebras: impala (N = 124 sightings), reedbuck (N = 47), elephant (N = 114), and waterbuck (N = 305). Bottom row shows species with similar body sizes to zebras: eland (N = 58 sightings), roan antelope (N = 28), hartebeest (N = 41), and topi (N = 172). Note that waterbuck also has a similar body size to zebras.

differ. I found that a greater proportion of sightings of zebras were made at far-off distances (>250 m) than were sightings of impala, waterbuck, or elephant but a smaller proportion were made at far-off distances than were sightings of reedbuck [4]. My observations therefore indicated that zebras were easier to spot at a distance than other herbivore species except in the case of reedbuck. This makes it difficult to argue that stripes conceal zebras at a distance during daylight hours.

In separate independent study in the same ecosystem, Waltert and colleagues (2008) determined sighting probabilities of large and midsized mammals during daytime foot surveys (Table 2.2). Of the 19 species reported, zebra had the fourth largest truncation distance (350 m)—i.e., the distance in meters in which the majority of the sightings were made. Therefore, the majority of zebra sightings on foot were made at distances farther than those of 15 other mammal species, many of which were gray-, dark brown–, or tan-colored species, again refuting crypticity as far as a different set of human viewers were concerned.

Sightings at dusk and dawn

At the start of the study, I wanted to spend long hours watching zebras in order to glean insights from their behavioral activities, so a team of us conducted observations at twilight to determine whether plains zebras are more cryptic than sympatric nonstriped species. This is an important time of day because this is when predators start to hunt. Three people viewed 10 sympatric species of large mammalian herbivores—buffalo, elephant, giraffe, hippopotamus, impala, reedbuck, topi, waterbuck, warthog, and zebra (wildebeest are absent from Katavi)—disappear in the evenings and appear in the mornings. Observers sat on the roof or hood of a Land Rover on the side of an open floodplain and used 8×30 binoculars to record whether they could or could not see various species every 5 min; different species were present on different days, but I only used sessions in which zebras were present with other species. Each observer made records in a personal notebook privately without discussion. Observations were conducted for a total of 26 mornings and 28 evenings by three observers (2005: Observer 1, 8 mornings and evenings; 2008: Observer 1, 9 mornings and 11 evenings; 2011: Observer 1, 9 mornings and evenings; Observer 2, 5 mornings and evenings; Observer 3, 4 mornings and evenings) and started before dawn (from 05.45 hrs until 08.00 hrs) or before dusk (18.00 hrs until 19.30 hrs). Evening observations were followed by preparing a fireside meal and sleeping in a tent next to the car or in the Land Rover itself, allowing us to get back up on the car before dawn the next morning.

The rate at which species vanished from sight in the evenings differed significantly, with zebras disappearing from sight (in the sense of being difficult to see) more slowly than any other species [5]. In the mornings, there were significant differences between the rates at which species came into view, with waterbuck appearing first, followed by zebras [6]. Often this was because waterbuck walked out of the woods from behind us onto the plain or else raised themselves from lying out in tall grass in the early mornings, so in reality it was zebras that were the first species that came into view. When zebras were compared with all the other nonstriped species combined, zebras disappeared significantly later than other species [7] and appeared significantly earlier [8], the latter effect being very marked (Figure 2.2). At great distances I could determine whether a gray spot was an animal rather than an inanimate object by the characteristic outline of its back. This was helped if there were several in a group because I could see a commonality to these shapes. Once I suspected it was an animal, even at

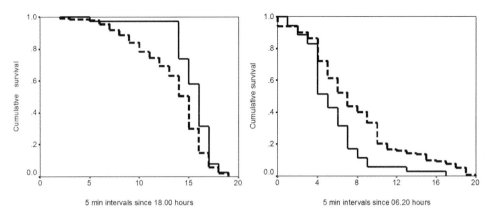

Figure 2.2. Kaplan-Meier curves showing probabilities of seeing herbivores of different pelage coloration. *Left:* Disappearance of striped zebras (continuous line) and 9 nonstriped herbivore species (buffalo, elephant, giraffe, hippopotamus, impala, reedbuck, topi, warthog, waterbuck) combined (dashed line) in the evenings (N = 171 events). Note, 1 on the y-axis, termed cumulative survival, denotes all individuals seen; 0 denotes none seen. *Right:* Appearance of striped zebras (continuous line) and 9 nonstriped herbivore species combined (dashed line) in the mornings (N = 159 events) as viewed from the side of floodplains by 3 observers. Note y-axis reversed in lower panel: here 1 denotes nothing seen, 0 denotes all seen.

a distance, I would recognize it as a zebra by its trapezoid shape, the characteristic shape of its head and thick neck, and by the frequent swishing of its tail (see section 5.2). Sizes of species as determined from their body weights or shoulder heights bore no relation to times of disappearances in the evenings averaged across days and observers [9]: zebras were the last to disappear irrespective of body weight. Similarly, average times of appearances in the mornings were unrelated to body size [10]; zebras were the second species to appear after waterbuck.

Unfortunately, there are difficulties with these sorts of observations. First, distance from the observer was not taken into account because groups were often farther away from observers than could be measured using a distance range finder, and groups moved during the course of a watch (Plate 2.1). Nonetheless, over the course of 482 scans in the evenings and 442 in the mornings, there was no obvious bias in distance that the 10 species were observed, save two: as noted, waterbuck often moved from the woodland situated immediately behind the observation vehicle onto the floodplain early in the morning, making them easy to see, and impala stayed relatively near the woodland fringe, making them more visible than might be expected. The second difficulty is that groups would sometimes walk on or off the plains during the course of a watch, causing them to "appear" or "disappear," respectively. Although these movements were uncommon, they add noise to the data. Consequently, I decided to use models and skins that could be set up at known distances from observers and would not move.

Out of plywood I constructed six two-dimensional life-size models cut into the shape of an "equid" (Plate 2.2) and set them up at equal distances from observers on the border of Katavi National Park ($\bar{X}$ = 158 m, range 65–288 m, depending on the day) in both wet and dry seasons over a series of mornings and evenings. The models were an adult plains zebra covered with black and white stripes using glossy paint (Plate 2.2A), and a herbivore painted gray using equal amounts of the same black and white glossy paint as the zebra model but mixed together (Plate 2.2B). A brown model was simply unpainted dark brown plywood with a glossy finish (Plate 2.2C). Gray and brown models acted as controls representing a "general sort of herbivore," such as a waterbuck and reedbuck, respectively. In addition, in one set of experiments I used a model with brown and white stripes instead of black stripes to determine whether black and white contrast was key to making models visible (Plate 2.2D). I also constructed two other models to test the effects of disruptive coloration (see section 2.2). Models were set up in a line about 5 m apart facing observers who stood near one another in a line and were asked to record whether they could see or not see each model every 5 min, using their naked eye, spectacles if they used them, then 8 × 30 binoculars, and then 10 × 50 binoculars. Observers made records in personal notebooks privately (Table 2.3; Plate 2.3).

Evidence was overwhelming that the zebra model was easier to see than the gray control and some evidence that it was easier to view than the brown control model. Observers noticed that the zebra model disappeared later on in the evenings than the gray model in wet season 2006 and in the dry seasons 2007 and 2008 but principally when using binoculars (Table 2.4). It came into sight earlier on in the mornings than the gray model in wet season 2006 and in the 2007 and 2008 dry seasons; a similar effect occurred with the naked eye and when wearing spectacles. With respect to the brown model, the zebra model disappeared later in the evenings in the 2008 wet and 2008 dry season, particularly when using binoculars, and appeared earlier in the morning with binoculars in the wet season 2008 and perhaps in the dry season 2007 (Table 2.4). In short, our independent observations indicated that (to human observers) black and

Table 2.3. Number of mornings and evenings in which different observers watched "equid" models and herbivore pelts appear and disappear, respectively, by season and year.

Number[a]	Models						Pelts			
	Z[b]	G	B	S	T	C	z1	z2	w	i
Wet 2005										
10	TL[c]			TL	TL	TL				
Wet 2006										
8	TC	TC			TC	TC				
Dry 2007										
8	TBJDE	TBJDE	TBJDE		TBJDE	TBJDE				
Dry 2008										
7	TBA	TBA	TBA				TBA	TBA	TBA	TBA
Wet 2008										
9	TJ	TJ	TJ				TJ[d]	TJ	TJ	TJ

[a]Refers to the total numbers of both mornings and evenings that models or pelts were watched.
[b]Z zebra, G gray control, B brown control, S striped brown, T tapir, C Cott model, z1 zebra skin 1, z2 zebra skin 2, w wildebeest skin, i impala skin.
[c]Each individual letter within the body of the table refers to the identity of different observers.
[d]2 mornings and 2 evenings only.

white striping is more conspicuous than gray or brown uniform coloration at times of low light intensity, a time when predators are active. Figure 2.3 provides such an example.

Unfortunately, these model experiments are somewhat artificial because the quality of light reflected off glossy paint and off zebra pelage differs, although the pattern of stripes on models was a good facsimile of a live zebra (see Caro and Melville 2012 for challenges in using models). For example, saturation on black stripes was greater for the zebra model than a zebra pelt as assessed by photographing both simultaneously at 400 ASA and importing the images into Photoshop 3 ($\bar{X}$ saturation of 10 black stripes = 25.6% and 12.7%, respectively), although it was more similar for white stripes ($\bar{X}$ saturation of 10 white stripes = 5.7% and 7.5%, respectively) and for measures of brightness ($\bar{X}$ brightness of 10 black stripes = 19.8% and 26.4%; $\bar{X}$ brightness of 10 white stripes = 76.7% and 83.3%, respectively). To be cautious, I therefore decided to use pelts of two zebras, a wildebeest, and an impala simultaneously along with my models in 2008. Pelts were hung on a clothesline at the same height and next to the plywood models (Plates 2.2C and 2.4). As I suspected from cursory observation, the painted zebra model was significantly easier to see than the two zebra skins in some situations [11, 12].

Yet despite my concerns, results using zebra pelts and a dark gray wildebeest pelt and a light brown impala skin were almost exactly the same

Table 2.4. Results of Cox's survival analyses with models as categorical variables, distance in meters as a continuous variable, and observer as a categorical variable.

	Evenings			Mornings		
	Wald χ^2	p-value	N	Wald χ^2	p-value	N
			Zebra model vs. gray model			
Wet 2006						
Naked eye			42	4.708[a,b]	0.030	64
Spectacles	5.575	0.018	63	10.896[b]	0.001	64
8 × 40 binoculars	26.112[a]	0.000	64	19.183	0.000	64
10 × 50 binoculars	19.876	0.000	64	17.781	0.000	60
Wet 2008						
Naked eye	[b]		110	3.726[b]	0.054[c]	112
Spectacles			56			56
8 × 40 binoculars	3.510	0.061	112			112
10 × 50 binoculars			112			112
Dry 2007						
Naked eye	[a,b]		190	[a,b]		190
Spectacles			40			150
8 × 40 binoculars	3.728[b]	0.054	189	11.366[a,b]	0.001	190
10 × 50 binoculars	9.059[b]	0.003	190	10.637[a,b]	0.001	178
Dry 2008						
Naked eye			144	14.775[a,b]	0.000	147
Spectacles	2.766	0.096	96	7.305[a,b]	0.007	98
8 × 40 binoculars	8.382	0.004	147	16.031[a]	0.000	147
10 × 50 binoculars	8.736	0.003	147	13.268[a]	0.000	147
			Zebra model vs. brown model			
Wet 2008						
Naked eye	[b]		110	[b]		112
Spectacles	2.976	0.084	56			56
8 × 40 binoculars	6.444	0.011	112	3.497	0.061	112
10 × 50 binoculars	6.003	0.014	112	3.967[a]	0.046	112
Dry 2007						
Naked eye	[a,b]		190	[a,b]		190
Spectacles			40			150
8 × 40 binoculars	[b]		189	3.150[a,b]	0.076	190
10 × 50 binoculars	3.147[b]	0.076	190	[a,b]		178
Dry 2008						
Naked eye			144	[a,b]		147
Spectacles			96	[a,b]		98
8 × 40 binoculars	3.665	0.056	147	[a]		147
10 × 50 binoculars	3.114	0.078	147	[a]		147

Note: Only significant results for comparisons of the zebra model with the gray and brown models are shown, df = 1 throughout; N refers to number of events (person X watches X disappearance or appearance). See text for details.

[a]Significant effect of distance from models.

[b]Significant effect of observer.

[c]Gray model appears before the zebra model.

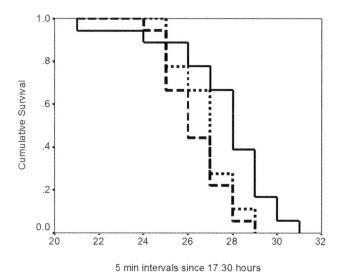

5 min intervals since 17.30 hours

Figure 2.3. Kaplan-Meier curves showing probabilities of observing a zebra model (continuous line), gray horse model (dotted line), and brown horse model (dashed line) through 8 × 30 binoculars during evenings (N = 112 events) in wet season 2008 as viewed by 2 observers. Note, 1 on the y-axis, termed cumulative survival, denotes all models seen; 0 denotes none seen.

as in the model experiments. One zebra pelt, call it z2, disappeared later than the wildebeest in the evenings in both wet and dry seasons, especially when binoculars were being used (Table 2.5), whereas the other pelt, z1, disappeared after the wildebeest in the wet season. In the mornings, both zebra pelts appeared before the wildebeest pelt came into view but for z2 in the dry season and z1 in the wet season. Regarding the smaller impala skin, there were normally no significant differences compared to the z2 pelt in the evenings, but the z1 pelt remained visible for longer. In the mornings z2 appeared earlier in the dry season and z1 earlier in the wet season than the impala skin when using low-powered binoculars. To summarize these findings using pelts, zebra coats stood out in comparison to dark gray pelts and light brown pelts, although less so in the latter case, and we did find variability in conspicuousness between zebra skins. Figure 2.4 provides an example.

Sightings at night

Next I felt I should extend these pelt experiments to nighttime, since this is when hunting is common (Hayward and Slotow 2009). Here I used a different protocol. I asked four people staying at our research camp to ob-

Table 2.5. Results of Cox's survival analyses with pelts as categorical variables, distance in meters as a continuous variable, and observer as a categorical variable.

	Evenings			Mornings		
	Wald χ^2	p-value	N	Wald χ^2	p-value	N
Zebra pelts vs. wildebeest pelt						
Wet 2008						
Naked eye	[b]		110	[a,b]		112
	5.573	0.018		3.691	0.005	
Spectacles			56			56
	4.129	0.042		3.408	0.065	
8 × 40 binoculars	3.873	0.049	112			112
	5.476	0.019		4.562	0.033	
10 × 50 binoculars	9.303	0.002	112	[b]		112
	7.939	0.005		[a]		
Dry 2008						
Naked eye	[b]		144	8.633[a,b]	0.003	147
				[a,b]		
Spectacles	2.698	0.100	96	2.809[a,b]	0.094	98
8 × 40 binoculars	3.697	0.055	147	3.601[a]	0.058	147
	4.101	0.043				
10 × 50 binoculars	3.109	0.078	147	[a]		147
Zebra pelts vs. impala pelt						
Wet 2008						
Naked eye	[b]		110	[a,b]		112
	2.855	0.091				
Spectacles			56			56
8 × 40 binoculars			112			112
	4.419	0.036		4.521	0.033	
10 × 50 binoculars			112	[b]		112
	4.062	0.044		[a]		
Dry 2008						
Naked eye	[b]		144	9.362[a,b]	0.002	147
				2.812	0.094	
Spectacles	3.088	0.079	96	4.182[a,b]	0.041	98
8 × 40 binoculars			147	2.776[a]	0.096	147
10 × 50 binoculars			147	[a]		147

Notes: Only significant results for comparisons of the two zebra pelts with the gray wildebeest pelt and brown impala pelt are shown, zebra pelt 2 shown directly above zebra pelt 1 in all cases, df = 1 throughout; N refers to number of events (person X watches X disappearance or appearance). See text for details.

[a]Significant effect of distance from pelts.

[b]Significant effect of observer.

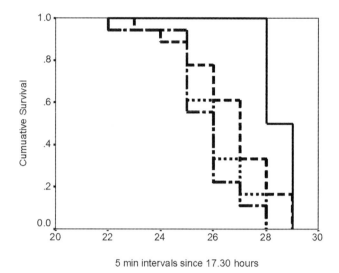

Figure 2.4. Kaplan-Meier curves showing probabilities of observing zebra pelts z1 (continuous line) and z2 (dashed line), impala pelt (dotted line), and wildebeest pelt (dashed and dotted line) through 8 × 30 binoculars during evenings (N = 112 events) in wet season 2008 as viewed by 2 observers. Note, 1 on the y-axis, termed cumulative survival, denotes all pelts seen; 0 denotes none seen.

serve pelts of different African ungulates briefly at night. Distances between skins and the viewing position were now short ($\bar{X}$ = 24.1 m, SE = 0.3). Specifically, I draped a wildebeest, a zebra, and an impala pelt over a clothesline or line of chairs at breast height between 20 and 26 m from the observer's chair (Plate 2.5). Pelts were set up within 3 m of one another at the same time but in a different arrangement each night and folded in a way to obscure their legs. Each observer was woken in the middle of the night and brought out of their bed separately. After one minute of sitting in the chair and adapting to the light, I asked each observer to identify whether they could see any pelts. Observers voluntarily reported whether they could definitely or possibly see anything (and I recorded them in this way) and if so point toward it. I later determined which skin this was from the direction in which they pointed. This experiment cost me some social capital in that my colleagues could see little point in being woken to observe animal pelts.

Observers found it difficult to see the pelts of the three species even at this close range (Table 2.6), especially when the moon was waxing. Nonetheless, there was a significant difference in the way in which observers reported definitely seeing the three species [13]: the impala was less likely

Table 2.6. Average percentage of occasions that 4 observers definitely saw, or possibly saw, or did not see pelts of three African ungulates (N = 52 person-nights).

	Zebra	Wildebeest	Impala
Definitely saw	32.8	20.0	3.7
Possibly saw	28.0	17.9	17.4
Did not see	39.3	62.2	78.9

Table 2.7. Number of nights in which only the zebra, wildebeest, or impala pelt was definitely or possibly seen. Also shown are number of nights when more than one pelt was definitely or possibly seen, and when none was seen (N = 52 person-nights).

Observer identity	A[a]	B	C	D
Only zebra	4	3	0	3
Only wildebeest	0	0	0	0
Only impala	0	1	0	0
More than one	6	5	4	7
None seen	4	7	5	3

[a]Refers to identity of four different observers.

to be seen than the zebra or wildebeest. It is noteworthy that the number of nights on which observers definitely saw or thought they saw only the zebra pelt was far greater than occasions for the other two pelts (Table 2.7). The likelihood that pelts of herbivores would be seen by observers was strongly influenced by distance for zebra [14], wildebeest [15], and impala [16] skins but not by observer identity. The probability of seeing the zebra or wildebeest pelt increased closer to full moon [14, 15]. Differences in visual acuity and color vision between humans and nonhumans notwithstanding (see below), this raises the possibility of predators being better able to notice these prey on cloudless nights around the full moon, but it is known that lions are actually less successful at hunting midsized prey on moonlit nights (van Orsdol 1984; Funston, Mills, and Biggs 2001). To summarize these nighttime observations: the zebra pelt was relatively easier to see than the other pelts, especially near full moon; the impala skin was difficult to spot on any night; and the wildebeest pelt appeared to observers on nights when the moon was large. In conclusion, for humans making observations at relatively close distances, zebras were more likely to be seen than impala and were often the only animal seen. These findings may bear some relevance for how zebras view their surroundings, because behavioral experiments indicate that domestic horses can see colors during the day and have similar thresholds of color vision to humans on moonlit nights (Roth, Balkenius, and Kelber 2007; Roth, Balkenius, and Kelber 2008).

2.2 Disruptive coloration

Some biologists have suggested that the sharply defined black and white stripes on a zebra prevent or delay recognition of the animal by drawing a predator's attention away from its true outline and concentrating attention on internal contrasts, allowing the animal to pass for part of the general environment (G. Thayer 1909; Stevens et al. 2006). This is called disruptive coloration, defined as "a set of markings that creates the appearance of false edges and boundaries and hinders the detection and recognition of an object's, or part of an object's, true outline and shape" (Stevens and Merilaita 2009b), and is another form of crypsis. Disruptive coloration has a number of subprinciples. These are differential blending, where at least some markings blend in with the background; maximum disruptive contrast, in which adjacent elements contrast strongly with each other and so draw the predator's eye to internal boundaries; disruptive marginal patterns, where markings touch the outline of the body, making parts of the animal appear continuous with the environment; disruption of the surface, where markings are placed away from the body's margins, thereby creating false edges that focus attention on internal objects that do not resemble the animal's outline; and coincident disruptive coloration, where markings cross over and join otherwise revealing body parts (Stevens and Merilaita 2009b). A common example of this last subprinciple is a black stripe passing through a black eye of an animal. In addition, Cott (1940) also suggested that irregular markings would be a component of disruptive coloration because variable shapes would give an impression of separate objects.

Some of these predictions, many of which are distinct from background matching, have been found to hold in the marine isopod *Idotea baltica* (Merilaita 1998). Moreover, experiments with artificial moth-like targets show that disruptive coloration does give prey protection from European bird predators, especially if some of the markings resemble the background (Cuthill et al. 2005; Schaefer and Stobbe 2006). Similar protective benefits have been found for artificial moths "attacked" by humans on computer screens (Fraser et al. 2007; Webster et al. 2013). Other experiments with birds have shown that disruptive coloration can be effective even when none of the markings resembles the background (Stevens et al.

2006), enabling disruptively colored species to occupy habitats that do not necessarily match the color of one of their markings.

For zebras to be disruptively colored, we should expect that their pattern of striping conforms to some of these subprinciples and, further, that they behave in ways that will maximize crypsis as described for background matching: namely, occupying backgrounds in which there are elements of black or white or both, living alone or in small groups, and freezing on sensing danger.

2.2.a Predictions

The design features of striping only partially conform to predictions of disruptive coloration (Table 2.8). In support, black and white stripes show sharp contrast with each other; indeed, this is one of the characteristics that make zebra coats so infamous. The vast majority of black and white stripes touch the body outline when viewed from any angle. (Note, how-

Table 2.8. Predictions of different characteristics of pattern elements of crypsis through disruptive coloration and observed characteristics for zebras.

Characteristics of pattern element	Prediction from disruptive coloration	Observed in zebras?
Maximum disruptive contrast	Adjacent stripes contrast strongly	Adjacent stripes do contrast strongly
Disruptive marginal patterns	More marginal pattern elements	Most stripes touch margins, but not on ventrum in all subspecies
Angle of elements	Touch body outline perpendicularly	Most stripes contact outline at right angles on haunch, legs, neck, and face but some obliquely on dorsum
Differential blending of colors with those in the environment	Some same as background	Possibly under certain conditions and in certain habitats
Shape	Complex and variable	Stripes simple and widths similar within restricted areas of the body
Regularity	Irregular	Regular on flank, rump, and forequarters
Disruption of surface	Elements placed away from body edge margins	Most stripes contact body edge
Coincident disruptive contrast	Markings cross over and join, otherwise revealing body parts	None on legs; stripes stop at eye or go around it

Source: Adapted from Merilaita (1998).

ever, that in Grevy's zebra and some subspecies of plains and mountain zebra, the lateral stripes taper out before meeting the ventrum, and in Grevy's zebra taper out near the top of the rump.) The vast majority of stripes meet the body's outline perpendicularly, although those above the rump run obliquely or nearly parallel to it. In some environments, such as dark forests, differential blending may occur, but in others it is unlikely to operate. Contrary to the hypothesis of disruptive coloration, however, most stripes consist of a simple pattern element—namely, straight bands—although a few bifurcate, especially above the foreleg and in the middle of the dorsum. Widths of adjacent stripes are similar, and are regular in spacing and orientation. Very few elements are internal to the body's margin. Finally, no markings cross over and link otherwise disjoint body parts or run through the eye, indicating an absence of coincident disruptive coloration. In short, the evidence for disruptive coloration is decidedly mixed.

2.2.b Sightings at dusk and dawn

In order to determine experimentally whether zebra striping might be a form of disruptive coloration, I compared a life-size zebra model with a brown model painted with similar white stripes; note that the latter model lacked black stripes and instead had brown stripes (Plate 2.2D). If the contrast between the black and white stripes serves to draw the eye away from the body outline, the zebra model should be more difficult to see. Second, I compared the zebra model with an equid painted to resemble a Malayan tapir (Plate 2.2E), a species that might be disruptively colored, although no formal coloration work has been done on this species. Malayan tapirs wear large blocks of black and white pelage—namely, a black head, black rump, and white torso. I used equivalent quantities of the same black and white paint on the "tapir horse" that I used to paint the zebra. I reasoned that if zebras are disruptively colored, they should be about as difficult to see as the "tapir horse." Finally, I compared the zebra model to a zebra painted with stripes running parallel to the animal's contour, exactly copying a drawing of a zebra in Cott's famous 1940 book (Plate 2.2F). In calling attention to the importance of contrasting elements being interrupted at or near the margin, Cott (1940) had drawn a "zebra" with the pattern of stripes running along the animal's contour, writing that such an individual would be easier to see than one in which the pattern cut across the contour (Figure 4.2 in Chapter 4). If the pattern of zebra stripes is a form of disruptive coloration, it should be more difficult to see than Cott's novel "zebra."

In a single set of observations conducted during the wet season 2005 involving two observers comparing the zebra and striped horse models, there was no significant difference in rates of disappearance of the two models in the evenings, or rates of appearance in the mornings, using the naked eye, spectacles or binoculars (Table 2.9). Therefore, based on these models, the sharp contrast between the black and white stripes does not make the model more difficult to see than the contrast between brown and white stripes.

In a series of three different sets of observations involving the zebra model and the "tapir horse," the tapir model appeared consistently earlier in the mornings, particularly when using the naked eye but also when observers wore spectacles (Table 2.9); this effect was absent when using binoculars. These findings are difficult to interpret because they suggest that zebra stripes are a more effective form of disruptive coloration than a "tapir horse" or that the latter model is a poor exemplar of disruptive coloration.

In a series of three experiments involving the zebra model and Cott's novel zebra, the two models did not differ in visibility, save in one instance: Cott's model appeared slightly earlier in the mornings in the wet season 2006 when observers used their naked eye (Table 2.9). In short, there was no strong evidence that Cott's model (a putatively nondisruptively colored model) was easier to see than the zebra model, hinting that the zebra model was not disruptively colored.

On the basis of these limited experiments with models, I am reasonably confident that the striping pattern of a zebra is not a form of disruptive coloration. The absence of differences between the zebra and Cott's model is most telling. Furthermore, no habitat occupied by the three zebra species has strong black and white elements, zebras live in groups, and they do not hide or freeze in response to predators, all of which speak against disruptive coloration.

2.3 Countershading

Countershading is a third mechanism by which animals can achieve crypsis (A. Thayer 1896; Poulton 1902; Cott 1940; Rowland 2008; Pennacchio, Lovell, et al. 2015). Countershading (usually dorsal pigmentary darkening) may result in self-shadow concealment, "where directional light, which would lead to the creation of shadows, is cancelled out by countershading" (Stevens and Merilaita 2009a), or, less likely in zebras, to obliterative shading, "where countershading leads to the obliteration of three dimen-

Table 2.9. Results of Cox's survival analyses with models as categorical variables, distance in meters as a continuous variable, and observer as a categorical variable.

	Evenings			Mornings		
	Wald χ^2	p-value	N	Wald χ^2	p-value	N
Zebra model vs. striped horse model						
Wet 2005						
Naked eye			39			40
Spectacles			80	[a]		80
8 × 40 binoculars	[a]		80	[a]		64
10 × 50 binoculars	[a]		80	[a]		55
Zebra model vs. tapir horse model						
Wet 2005						
Naked eye			39			40
Spectacles			80	3.695[a]	0.046	80
8 × 40 binoculars	[a]		80	[a]		64
10 × 50 binoculars	[a]		80	[a]		55
Wet 2006						
Naked eye			42	3.055[a,b]	0.081	64
Spectacles			63	5.046[b]	0.025	64
8 × 40 binoculars	[a]		64			64
10 × 50 binoculars			64			60
Dry 2007						
Naked eye	5.251[a,b]	0.022	190	9.819[a,b]	0.002	190
Spectacles			40			150
8 × 40 binoculars	[b]		189	[a,b]		190
10 × 50 binoculars	[b]		190	[a,b]		178
Zebra model vs. Cott zebra model						
Wet 2005						
Naked eye			39			40
Spectacles			80	[a]		80
8 × 40 binoculars	[a]		80	[a]		64
10 × 50 binoculars	[a]		80	[a]		55
Wet 2006						
Naked eye			42	2.872[a,b]	0.090	64
Spectacles			63	[b]		64
8 × 40 binoculars	[a]		64			64
10 × 50 binoculars			64			64
Dry 2007						
Naked eye	[a,b]		190	[a,b]		190
Spectacles			40			150
8 × 40 binoculars	[b]		189	[a,b]		190
10 × 50 binoculars	[b]		190	[a,b]		178

Note: Only significant results for comparisons of the zebra model with the striped horse, tapir horse, and Cott's zebra models are shown, df = 1 throughout, N refers to number of events (person X watches X disappearance or appearance). See text for details.
[a]Significant effect of distance from models.   [b]Significant effect of observer.

sional form" (Stevens and Merilaita 2009a). Zebras are countershaded in two ways: by the black flank stripes failing to reach the ventrum, as in Grevy's zebra, or by black flank stripes becoming thinner from dorsum to ventrum, thereby exhibiting more white pelage. Interestingly, Mottram (1915) specifically discussed this type of pattern of thinning soon after A. Thayer introduced the idea of graded background coat colors in 1896.

Amanda Izzo scored up to 12 individual photographs of equid subspecies (see section 8.2 and Appendix 5) in two ways. First, she scored the extent of belly striping as follows: ventrum white (0), stripes not reaching the ventral midline, or faint stripes (1), and stripes definitely reaching the ventral midline (2). Table 2.10 shows that Grevy's zebras have no belly stripes, mountain zebras have some, and plains zebras have heavily striped ventra. None of the asses or Przewalski's horse has belly stripes. Second, she noted whether there was a mismatch between belly and body color (termed countershading here). Phylogenetic analyses of this second measure at both subspecific and species levels using a yes or no criterion revealed that countershading is no more prevalent than absence of countershading in unforgiving sunny habitats. Specifically, the geographic ranges of equid species and their subspecies do not show high overlap with parts of the Old World where maximum temperatures reach 30°C, or with open habitat [17]. In both of these environments, the sun is likely to cast a strong

Table 2.10. Mean belly stripe scores for subspecies of zebras (see Chapter 8).

Species Subspecies	N	Score*
Grevy's zebra		
E. grevyi	11	0
Mountain zebra		
E. z. hartmannae	10	0.75
E. z. zebra	9	1.00
Plains zebra		
E. b. antiquorum	12	1.00
E. b. burchellii	10	0.92
E. b. chapmannii	11	1.00
E. b. crawshayi	7	1.00
E. b. zambeziensis	8	1.00
E. b. boehmi	11	1.00
E. b. quagga	12	0

*0 denotes no belly stripes with the ventrum being white; 1 denotes stripes not reaching the ventral midline, or faint; 2 denotes stripes definitely reaching the ventral midline (A. Izzo, unpublished data).

shadow on the body. Somewhat surprisingly, then, and contrary to some anecdotes (Moodley and Harley 2006), lighter ventra in equids are not associated with these sunny environments as they are for other herbivore species (Stoner, Caro, and Graham 2003; Allen et al. 2012).

2.4 Zebras as seen by nonhumans

Unfortunately, we have little idea of how predators or even conspecific zebras view striped coats under natural situations. This is a potentially serious problem because humans have anomalously acute photopic (daytime) spatial vision that far surpasses the ability of nonprimate mammals to resolve detail, and our diurnal activity may lead us to overestimate the conspicuity of striped pelage (Uhlrich, Essock, and Lehmkuhle 1981). To understand how nonhumans view zebra stripes, Amanda Melin, Donald Cline, Chihiro Hiramatsu, and I estimated the maximum distance at which stripes on the three species of zebras can be resolved by different species of observer using spatial-vision-based digital image filtering techniques to compare images of zebras and other prey as they might be seen under high, intermediate, or low light conditions (Melin et al. 2016).

To calculate the maximum distance at which the thinnest and widest stripes on the body of adult plains, mountain, and Grevy's zebras could be resolved by different predators, other zebras, and humans, we needed to measure the luminance of black stripes and white stripes of live captive zebras to quantify the mean Michelson luminance contrast

$$(L_{max} - L_{min})/(L_{max} + L_{min}),$$

where L is luminance, under a range of luminance conditions from daylight to twilight. Luminance of zebra stripes and the immediate surrounding area (sky, grass, dirt) above, below, to the left, and to the right of the animal was recorded by Amanda Melin using an LS-110 Minolta Spot Photometer with a one-degree acceptance angle from live Grevy's zebras at the Calgary Zoo at 14.50 hrs under full sun. Luminance under photopic (sun with light cloud) and then mesopic (twilight) conditions was measured from 19.50 hrs past sunset until ambient levels fell below the range of the meter at 20.43 hrs. Zebras were measured in both open (full sun) and shaded (enclosure structures, trees) conditions. In the former case the luminance of stripes both above and below the natural body shadow were measured. Absolute luminance was higher in stripes exposed to full sun; however, luminance contrast was highly conserved across conditions.

Stripe width measurements were taken from a total of nine preserved study pelts by Hannah Walker.

In a parallel part of the study that allowed us to visualize how non-humans see ungulates, I took photographs of adult and subadult plains zebras, waterbuck, topi, and impala; pelts of plains zebras, wildebeest, and impala; and of the life-size plywood zebra model and model horses in Katavi National Park in August and September 2012. Photographs were taken with a 300 mm zoom 167 telephoto lens mounted on a Nikon D50 digital camera set at 70 mm in manual mode with EV steps set at 1/3rd 168 as a default, on autofocus, and shot in RAW. White balance was set in the field prior to each session. Objects (animals, pelts, or models) were photographed from 100–200 m away; precise distances were subsequently recorded using a range finder. Within less than 30 sec after the photograph was taken, I alighted from the vehicle and went to the object's location and erected a color checker card 1 m above ground. I then returned to the vehicle and took another photograph of the checker card with the same zoom setting (70 mm) using exactly the same aperture and shutter speed settings. (I could not do this before taking the first photograph as I would frighten the zebras away.) Due to digital photography limitations, the photographs were taken under a range of high mesopic twilight through low to high photopic daylight conditions.

Methods of deriving contrast sensitivity functions (CSFs) and simulating spatial, color, and luminance vision differences involving linear analysis of digital photographs can be found in Melin et al. (2016).

We found that humans can resolve stripes at considerably greater distances than can zebras or predators under daylight and dusk conditions but that all species had roughly equivalent resolution on moonless nights (scotopic conditions). Specifically, under photopic daylight conditions, humans can resolve stripes at 2.6 times greater distances than zebras can, and 4.5 and 7.5 the distances that lions and spotted hyenas, respectively, can (Table 2.11). Wide stripes might therefore be a signal to lions at only 44–98 m or less, depending on the zebra species, and to hyenas at 26–59 m or less, again depending on the zebra; narrow stripes would be discernible only at far shorter distances (Table 2.12). At dusk, however, when hunting often begins, humans can still resolve stripes farther away than can nonhumans: 1.9 times farther than zebras, 3 times farther than lions, and 5.1 times farther than hyenas. Under these mesopic conditions, lions can discern stripes at 25–56 m or less and hyenas at 15–34 m or less, but beyond these distances zebras will appear gray. Under scotopic conditions when hunting also occurs, humans can resolve stripes at 0.8 times the distance

Table 2.11. Relative performance of different species under different lighting conditions.

		Human best 20/10	Zebra	Lion
Daylight	Human 20/10			
	Zebra	2.6[a]		
	Lion	4.5	1.7	
	Hyena	7.5	2.9	1.7
Dusk	Human 20/10			
	Zebra	1.9		
	Lion	3.0	1.6	
	Hyena	5.1	2.7	1.7
Moonless night	Human 20/10			
	Zebra	0.8		
	Lion	0.7	0.9	
	Hyena	1.1	1.4	1.4

Source: From Melin et al. (2016).
[a]Refers to how organism in the column can see zebra stripes compared to that in the row. Thus, humans can see zebra stripes 2.6 times farther away than can other zebras.

that zebras can, 0.7 times that of lions, and 1.1 times the distance of hyenas. For a lion, it can resolve the widest stripes at 6–14 m away, and a hyena can at only 4–9 m away (Melin et al. 2016); and far less for the narrowest stripes. As an illustration, Plate 2.6 shows the effects of decreasing ambient light when viewed at 16 m away by humans and lions under photopic (*A, B*), mesopic (*C, D*), and scotopic (*E, F*) conditions.

If striping is a form of crypsis, zebras should be more difficult to see through a predator's eye than a uniformly colored herbivore of similar size, but our analyses showed otherwise. For example, in the top set of panels in Plate 2.7, it can be seen that stripes on individual zebras do not make zebras blend together in simulated lion vision; individual zebras can still be clearly discriminated from one another. In the pair of panels second from the top, even the more distant zebras are as easy to discern as topi. Comparison of the top three panels indicates that waterbuck, topi, and zebras are all relatively easy to see. Finally, from a lion's point of view, the smaller impala are more difficult to detect in the tall grass present in the scene.

To assess the conspicuity of zebra stripes against a background of trees, we used photographs of live zebras and also life-size models of a striped zebra and a gray horse model in woodland. A solitary zebra appears less conspicuous in a forested setting than on the open plains, and a striped zebra model appears less conspicuous than a solid gray model in woodland to all visual systems that we simulated (Melin et al. 2016). This effect may in part be due to increased luminance contrast outside of wooded areas:

Table 2.12. Estimated maximum distances in meters at which zebra stripes can be resolved by different species by body region.

Relative size	Position	Average width (cm)	Human	Zebra	Lion	Hyena
PHOTOPIC						
Plains zebra (N = 5 pelts)						
Widest	flank	5.23	360	140	80	48
Narrowest	forelimb	1.19	82	32	18	11
Mountain zebra (N = 1)						
Widest	neck	6.38	439	170	98	59
Narrowest	forelimb	1.41	97	38	22	13
Grevy's zebra (N = 3)						
Widest	rump	2.83	195	76	44	26
Narrowest	forelimb	1.00	69	27	15	9
MESOPIC						
Plains zebra (N = 5)						
Widest	flank	5.23	139	75	46	28
Narrowest	forelimb	1.19	32	17	10	6
Mountain zebra (N = 1)						
Widest	neck	6.38	169	91	56	34
Narrowest	forelimb	1.41	37	20	12	7
Grevy's zebra (N = 3)						
Widest	rump	2.83	75	40	25	15
Narrowest	forelimb	1.00	27	14	9	5
SCOTOPIC						
Plains zebra (N = 5)						
Widest	flank	5.23	8	10	11	7
Narrowest	forelimb	1.19	2	2	3	2
Mountain zebra (N = 1)						
Widest	neck	6.38	9	12	14	9
Narrowest	forelimb	1.41	2	3	3	2
Grevy's zebra (N = 3)						
Widest	rump	2.83	4	5	6	4
Narrowest	forelimb	1.00	1	2	2	1

Source: From Melin et al. (2016).

measures of both live zebras in captivity and photographs of wild zebras reveal greater luminance contrast between animals and unobstructed sky on the plains than between animals and vegetative backdrops such as that found in woodland.

Our findings certainly show that striping is unlikely to be a form of crypsis at a distance because predators would not be able to discern the black stripes and see them blend in with trees behind or see white stripes blend in with bright shafts of light between trees (background matching). This is because predators could not discern stripes beyond a maximum distance of 59 m (hyenas) or 98 m (lions) in daylight. For the same reasons stripes could not break up the zebra's outline through disruptive coloration. At dusk, the distances at which stripes can be resolved reduces to 34–56 m, depending on which predator, and to just 9–14 m on moonless nights—distances at which predators could likely hear zebras moving or breathing!

Comparisons between uniformly colored species and plains zebras seen on the plains (Plate 2.7) show that all species are conspicuous because their body silhouette is high in luminance contrast relative to the sky. Thus, in open plains environments, a setting where zebras spend much of the year, stripes do not disrupt the outline of the body and so are unable to confer cryptic advantages compared to unstriped species. In some wooded environments, where stripes might resemble branches and tree trunks of saplings, again stripes can only be resolved by predators nearby; zebras appear gray under other conditions. Only a stationary, silent zebra at close range in a woodland habitat could incur any benefits of crypsis due to stripes, but at this distance scent or noise would be a cue to predators. In short, Melin and colleagues' analyses do not support stripes conferring a form of crypsis against predators under any of a variety of conditions.

2.5 Conclusions

Evidence presented in this chapter lends no support to striping being a form of crypsis either through background matching or disruptive coloration (Table 2.13). In brief, the three species of zebras feed on grasses in open habitats that include open plains, treeless montane grasslands, semideserts, and lightly forested woodlands. Some populations occupy plains habitat for some of the year and woodland areas for others. It is difficult to imagine one type of coloration blending in well with all of these locales.

Zebras are social animals; all three species may be found in large group sizes during portions of the year. Mountain zebras and plains zebras live in harems of approximately six individuals but form temporary aggregations; Grevy's zebra group sizes range from single males or females with young at hoof to large temporary herds in the wet season. As expected from their social structure, zebras are not quiet, slow-moving, secretive species that freeze upon seeing a predator, behavior that often goes hand in hand with cryptic pelage. Instead, they group together and confront or actively defend themselves from predators.

Comparative data show that striped and nonstriped equids differ little in their habits: they all live in open environments, in medium- to large-size groups, and do not try to hide from predators. My observations show that plains zebras can be seen at considerable distances compared to other

Table 2.13. Summary of findings regarding crypsis.

Background matching unlikely because zebras:

Are often found on the plains rather than in woodlands.
Are found in medium to large groups.
Do not rely on immobility.

Humans
Zebras can be seen at considerable distances.
Zebras disappear later in the evenings compared to live heterospecifics, models, and pelts of other
    species.
Zebras appear earlier in the morning compared to live heterospecifics, models, and pelts of other
    species.
Zebra pelts easier to see at night than those of other species.

Predators
Large predators can only discern stripes close up.

Disruptive coloration unlikely.

Design features only partially support this idea.

Humans
Zebra model disappears at same rate as nondisruptively and disruptively colored models.
Zebra model appears more slowly than disruptively colored model and at same rate as nondisruptively
    colored models.

Predators
Large predators can only discern stripes close up.

Countershading unlikely.

Not associated with hot, open areas.

---

sympatric ungulates of similar group sizes and body sizes. Like Roosevelt and Heller (1914, p. 685), who wrote, "The zebra frequently whisks its tail which at once attracts attention," or Selous (1908, p. 22), "The whisking of their tails will probably be the first thing to catch one's eye," I found that this movement alerted me to their presence. To the human eye, zebras in their natural habitat disappear later in the evenings and appear earlier in the mornings than other species of ungulate. Experiments with models that eliminate nonstandardized distances and inability to control animal movements also show that zebra models disappear later in the evenings and appear earlier in the mornings than classical cryptically colored models. Experiments with herbivore skins confirm that zebras are easier to see than wildebeest or impala skins at dusk, at night, and at dawn. Thus, stripes offer no cryptic advantages through background matching.

I have less data on crypsis through disruptive coloration. Nonetheless, some of the arguments against background matching can be borrowed. For example, zebras aggregate in large groups, do not rely on immobility, and are easy to spot at a distance. Experiments show that a zebra model appears more slowly in the mornings than a putatively disruptively colored model and is as easy for people to see at dusk and dawn as nondisruptively colored models. The design features of stripes are further evidence against striping disrupting the outline of the zebra's body or drawing a predator's attention away from the outline.

Countershading, which in zebras consists of an absence of striping on the belly, does not appear to be a form of crypsis, insofar as it is not associated with open habitats or very hot climates. Countershading is principally seen in smaller ruminants living in well-lit environments and close to the equator, where their countershading is near to the optimal expected for self-shadow concealment (Allen et al. 2012). Zebras are large and thus relatively conspicuous and may not be able to benefit from crypsis afforded by countershading.

Analyses using digital images of zebras and other prey passed through specific spatial and color filters to simulate their appearance for predator visual systems show that stripes are difficult to resolve except at surprisingly close distances and thus cannot confer crypsis through background matching or disrupting the zebra's outline. Zebra profiles are more difficult to detect in woodlands, and stripes confer a minor advantage over solid pelage in masking body shape, but this effect is stronger for humans than for predators. Overwhelmingly, stripes do not appear to confer antipredation advantages for their bearers and cast serious doubt on their being a form of camouflage. It is worth remembering Selous's (1908, p. 22) observation: "As, however, zebras have a very strong smell, and lions usually hunt them by scent and at night, I cannot think that their coloration, whether it be conspicuous or not matters very much to them."

Finally, if zebra stripes were a form of crypsis, it raises the question of why similar contrasting stripes are absent in other African herbivores such as wildebeest and topi that live in savannahs and are predated by the same species of large carnivore. In some woodland artiodactyl species, such as bongo and bushbuck, pelage striping likely serves as background matching (Stoner, Caro, and Graham 2003), but it is far less distinct than that of zebras. Striping more reminiscent of zebras is found in the zebra duiker, but again the stripes show less contrast. Our working hypothesis should be that different forms of striping serve different functions across herbivores.

Statistical tests

[1] Number of zebra sightings made in closed habitats (W1, W2, W3, and B combined; see Table 2.1) (48) vs. open habitats (BG, WG, and G combined) (166) in

Katavi National Park in all seasons against that expected on the basis of habitat availability, χ^2 = 0.141, df = 1, NS (NS, not significant, refers to all p values >0.1 throughout the book).

[2] Number of zebra sightings made in closed habitats (W1, W2, W3, and B combined) (5) vs. open habitats (BG, WG, and G combined) (73) in Katavi National Park in the dry season (September–November 1995, August–October 1996) was significantly less than that expected on the basis of habitat availability, χ^2 = 12.010, df = 1, p < 0.0001.

Number of zebra sightings made in closed habitats (W1, W2, W3, and B combined) (19) vs. open habitats (BG, WG, and G combined) (66) in the intermediate season (December 1995; May, June, and November 1996) against that expected on the basis of habitat availability, χ^2 = 1.142, df = 1, NS.

Number of zebra sightings made in closed habitats (W1, W2, W3, and B combined) (24) vs. open habitats (BG, WG, and G combined) (27) in the wet season (January–March, December 1996) was significantly greater than that expected on the basis of habitat availability, χ^2 = 34.931, df = 1, p < 0.0001.

[3] Rank average sighting distance vs. rank body weight controlling for rank average group size seen on Katavi transects, Pearson correlation on ranked data, N = 16 species, r = 0.264, NS; rank average sighting distance vs. rank average group size controlling for rank body weight, N = 16 species, r = 0.487, p = 0.040.

[4] Number of sightings of zebra made between 0–250 m vs. 251–500 m compared to impala χ^2 = 21.950, df = 1, p < 0.001; reedbuck χ^2 = 26.753, df = 1, p < 0.001; waterbuck χ^2 = 3.851, df = 1, p < 0.1; elephant χ^2 = 17.079, df = 1, p < 0.001.

[5] Disappearance of herbivores in the evenings: N = 171 events, species Wald = 42.858, df = 9, p < 0.0001; person Wald = 3.080, df = 2, NS.

[6] Appearance of herbivores in the mornings: N = 159 events, species Wald = 31.160, df = 9, p < 0.0001; person Wald = 1.757, df = 2, NS.

[7] Disappearance of striped and nonstriped species of herbivore in the evenings: N = 171 events, species Wald = 5.841, df = 1, p = 0.016; person Wald = 3.208, df = 2, NS.

[8] Appearance of striped and nonstriped species of herbivore in the mornings: N = 159 events, species Wald = 7.547, df = 1, p = 0.006; person Wald = 2.162, df = 2, NS.

[9] Average time that 10 species of herbivore disappeared in the evenings against body weights, r_s = 0.442, NS; against shoulder heights, r_s = 0.212, NS.

[10] Average time that 10 species of herbivore appeared in the mornings against body weights, r_s = –0.418, NS; against shoulder heights, r_s = –0.716, NS.

[11] Evening watches in which the zebra model disappeared significantly later than a zebra pelt: Wet season 2008, 8 × 30 binoculars, N = 112 events, z2 (zebra pelt #2) Wald = 3.036, df = 1, p = 0.081. Dry season 2008, N = 147 events, 10 × 50 binoculars, z1 (zebra pelt #1) Wald = 4.626, df = 1, p = 0.031, z2 Wald = 4.170, df = 1, p = 0.041.

[12] Morning watches in which the zebra model appeared significantly earlier than a zebra pelt: dry season 2008, N = 147 events, naked eye, z1 Wald = 11.345, df = 1,

p = 0.001, z2 Wald = 4.067, df = 1, p = 0.044; spectacles, N = 98 events, z1 Wald = 5.319, df = 1, p = 0.021; 8 × 30 binoculars, N = 147 events, z1 Wald = 4.837, df = 1, p = 0.028, z2 Wald = 2.860, df = 1, p = 0.091; 10 × 50 binoculars, N = 147 events, z1 Wald = 5.699, df = 1, p = 0.017, z2 Wald = 4.519, df = 1, p = 0.034.

[13] Number of nights that four observers could definitely see a zebra pelt, wildebeest pelt, and impala pelt at night: Kruskal-Wallis test, χ^2 = 7.758, df = 2, p = 0.021. Comparing zebra and wildebeest, Mann-Whitney U test, U = 5.0, NS; zebra and impala, U = 0, p = 0.020; wildebeest and impala, U = 0, p = 0.020.

[14] Likelihood of observers saying they definitely saw a zebra pelt as a function of distance: N = 52 events, Wald = 8.181, df = 1, p = 0.004; observer ID, Wald = 3.002, df = 3, NS; and moon phase, Wald = 4.979, df = 1, p = 0.026 using a binary logistic regression.

[15] Likelihood of observers saying they definitely saw a wildebeest pelt as a function of distance, N = 52 events, Wald = 10.788, df = 1, p = 0.001; observer ID, Wald = 3.361, df = 3, NS; and moon phase, Wald = 6.210, df = 1, p = 0.013 using a binary logistic regression.

[16] Likelihood of observers saying they definitely saw an impala pelt as a function of distance, N = 52 events, Wald = 7.057, df = 1, p = 0.008; observer ID, Wald = 1.028, df = 3, NS; and moon phase, Wald = 0.832 df = 1, NS using a binary logistic regression.

[17] Equid subspecies (N = 20): phylogenetic logistic regressions at the 0/1 level: average annual maximum temperature of 30°, NS; open habitat, NS. Equid species (N = 7): average annual maximum temperature of 30°, NS; open habitat, NS.

Figure 3.0. Lioness attacking an adult plains zebra. Most lion predation attempts involve a short chase, climbing on to the zebra's rump or back, and using weight to topple the prey over. (Drawing by Sheila Girling.)

THREE

Predation and aposematism

3.1 Aposematism in mammals

In nature bright contrasting achromatic (without color) and chromatic (colorful) external appearances sometimes signal animal defenses (Ruxton, Speed, and Sherratt 2004), and zebras have this sort of appearance. General books about animal biology have made this association (e.g., Louw 1993). For example, Matthews (1971, p. 350) wrote, "The curved and angled stripes on the sides of the face may well draw attention to the mouth—and zebras bite." So are the unusual stripes of zebras a type of warning coloration or aposematism, defined as "an appearance that warns off enemies because it denotes something unpleasant or dangerous" (Poulton 1890, p. 340)? Although Wallace was the first to recognize that conspicuousness might be a warning signal (Wallace 1867b; Wallace 1896), Poulton (1890) explored its component features more fully, suggesting that aposematism was associated with two or more contrasting colors within a prey's pattern; unpleasant qualities that may be concentrated in conspicuous parts; conspicuous behaviors, including an absence of hiding or escape behaviors; sluggishness; aggregation; and even simply being freely exposed rather than hidden. Stevens and Ruxton (2012) highlighted additional features: internal contrast between markings on the body and contrast with the background; repeated pattern elements; distinctiveness from undefended species; and red, yellow, and black markings. Superficially, at least, zebra coats fit many of these criteria.

Before proceeding further, it is worth considering the

natural history of warning coloration. Aposematism is relatively common among invertebrates and poikilothermic (cold-blooded) vertebrates, where it often takes the form of black, yellow, or red color patches juxtaposed against each other; however, it is much less common in homeotherms, being found in only a few mammals (Caro 2013) and unrelated bird groups (Cott 1946–1947; Baker and Parker 1979; Gotmark 1994; Dumbacher and Fleischer 2001). In carnivores it is principally observed in small to midsize plantigrade species that walk with their toes and metatarsals (five long bones in the foot) flat on the ground, that have a stocky build and are unable to run fast, and that live in open habitats where opportunities for escape through climbing trees or hiding among rocks are limited (Stankowich, Caro, and Cox 2011). Skunks are the best-known group of aposematic mammals, but warning coloration is also exhibited by other midsize carnivores such as badgers and civets (Caro 2009). In all these species, blazes of black and white pelage are juxtaposed, and this form of coloration is associated with noxious anal secretions that can in some species be directed in a stream against potential attackers (Lariviere and Messier 1996; Stankowich, Caro, and Cox 2011). Other aposematic mammals have black, white, and brown bands of varying degrees of conspicuousness that adorn their quills or spines (see Caro 2005). These include the echidnas (monotremes), which use their spines to prevent them from being dislodged from crevices; hedgehogs, which roll into a ball or jump toward their opponent; tenrecs (insectivores), which hiss and head buck and can drive detachable neck spines into a predator; and Old and New World porcupines (rodents), which rattle their spines or clack their teeth or stamp and may run backward into an intruder.

Of course, the most notable characteristic of a zebra when viewed up close by humans is its conspicuous repeated black and white pattern elements with high luminance contrast within the coat, and the coat standing out from the background (see Roosevelt and Heller 1914) (Table 3.1). Stripes can be resolved at close distances by predators too (section 2.4). In other contexts, these traits have been shown to improve avoidance

Table 3.1. Common aposematic traits and their occurrence in zebras.

Aposematic traits	Comments regarding zebras
Repeated pattern elements	Stripes repeated regularly along sections of the body.
Contrasting internal markings	Black and white stripes juxtaposed.
Contrasting coloration with background	Generally confirmed with humans when viewed up close, but less so for carnivores.
Black and white markings	Similar to other aposematic mammals.

learning by predators (Sillen-Tullberg 1985; Hauglund, Hagen, and Lampe 2006) and to help predators detect artificial prey, learn prey characteristics, and forget prey less easily (Gittleman and Harvey 1980; Roper and Redston 1987; Roper 1994; Alatalo and Mappes 1996; Ruxton, Sherratt, and Speed 2004). Furthermore, conspicuousness based on brightness contrast rather than varying hues does not demand that observers be trichromats (i.e., seeing in three colors) (Prudic, Skemp, and Papaj 2006), so it could be effective in deterring visually dichromatic mammalian carnivores. It is possible, then, that zebra coloration might be a warning signal to certain predators when viewed up close.

The defensive component of aposematism is more problematic. Zebras do not manufacture toxic secretions, do not have armored defenses, and are palatable, so, by elimination, could only be advertising unprofitability (Leimar, Enquist, and Sillen-Tullberg 1986). Unprofitability might stem from forceful kicks, strong bites, a tough epidermis, ferocious disposition, or perhaps large body size (Edmunds 1974). Certainly, equids are renowned for their bites (Matthews 1971) and powerful kicks, but the extent to which these are dangerous to predators or more dangerous than other similar-size herbivores is not really known. Alternatively, unprofitability might derive from flight speed: in some insects aposematism appears to signal particularly good escape capability (V. Thompson 1973; D. Gibson 1974; D. Gibson 1980). Nonetheless, qualitative observations of antipredator behavior, all that are available at present, show no marked differences between striped and unstriped equids (Table 3.2).

If black and white striping in zebras is a form of aposematism, we might expect that zebras would face similar ecological challenges as other aposematic mammals. These include intense predation pressure (Merilaita and Kaitala 2002), nocturnal predation (Stankowich, Havercamp, and Caro 2014), and alternative palatable prey being available (Sherratt 2003). These conditions do pertain in many parts of the ranges of the three zebra species. For example, in Serengeti 30% of plains zebra annual mortality is attributed to predation (Grange et al. 2004), most of the predation by lions is nocturnal (Scheel 1993), and alternative ungulate prey are available in large numbers in most protected areas. The aposematic hypothesis therefore demands serious consideration.

3.2 Signaling component of aposematism

If zebras are aposematic, we would expect them to be distinctive close up (Merilaita and Ruxton 2007), make no attempt at being cryptic (Sherratt

Table 3.2. Qualitative descriptions of antipredator behavior in equids.

Species	Observations
Plains zebra	Family group stallion defends by biting or kicking, barking.[a]
	Fleeing behavior, stallion defends mares, mares defend foals. Biting and kicking hyenas.[b]
	Flailing, biting, standing rigid, and bunching in response to lions; lowered head as if to bite and fleeing behavior in response to wild dogs.[c]
	Joining to form groups with stallions in rear kicking and biting. Guards stay awake at night. Stallion lunging and baring teeth in response to wild dogs. After mare and foals were separated, ten zebras came back and rescued them. Biting in response to hyena.[d]
	Lunging with bared teeth and ears back along neck in response to wild dogs.[e]
Mountain zebra	A male kicked a hyena to death.[f]
	Mare leads herd away, stallion stays in defense at rear, kicks.[a]
Grevy's zebra	Approaching in erect posture, braying, hazing, biting, tail swish, ears back, rearing, kicking.[g]
African wild ass	No information.
Asiatic wild ass	Stallions help defend offspring year round; stallions from different families chase wolves cooperatively.[h]
Kiang	Vigilance.[i]
	Head up, nostrils flared, pawing ground, circling, rearing on hind legs, and fleeing in response to man.[j]
Przewalski's horse	Mare kicks backward and sprinkles urine into attacker's eyes with hoofs; stallion bites and cuts with forelegs.[k]
	Lookout makes noise at sign of danger. Group flees with stallion at rear biting and kicking individuals who are not moving fast enough.[l]

[a]Skinner and Chimimba (2005).	[d]van Lawick-Goodall and van Lawick-Goodall (1970).	[f]Estes (1993).	[j]Groves (1974).
[b]Kruuk (1972).		[g]Estes (1991).	[k]Bökönyi (1974).
[c]Schaller (1972).	[e]van Lawick (1973).	[h]Feh et al. (2001).	[l]Mohr (1971).
		[i]Xu et al. (2013).	

2002), possess multimodal signals as is commonplace in unprofitable prey (Rowe 1999; Rowe 2002), live in exposed habitats (Stankowich 2012), and perhaps aggregate (Alatalo and Mappes 1996; Rippi et al. 2001). Informally, these predictions are met. Stripes can be resolved by lions between 44 m and 98 m in daylight and by spotted hyenas between 26 m and 59 m (Table 2.12); zebras make no attempt at being cryptic by hiding, skulking, or moving surreptitiously; they are reportedly noisy as well as being visually conspicuous (Roosevelt and Heller 1914; Morris 1990); Grevy's and mountain zebras principally occupy exposed habitats (Table 3.3); and all species form temporary feeding aggregations (Table 1.3). Under threat of predation, they can be bold, at least toward certain predators, and defend themselves actively (Table 3.2). We might also expect them to show signs of surviving predation attempts.

Table 3.3. Percentage vegetation overlap with seven vegetation types in Africa.

	Plains zebra	Mountain zebra	Grevy's zebra
Arid desert and semidesert	6.0	43.2	63.7
Savannah and steppe	74.2	35.1	30.8
Mediterranean	0	21.6	0
Dry forest	11.2	0	0
Rain forest	7.6	0	5.5

Source: A. Izzo, unpublished data.

3.2.a Visibility

As described in section 2.1.b, I noted the distances at which I observed groups of zebras and other large and midsize mammals from a vehicle during monthly transects driven in and around Katavi National Park. Here I divide estimated sighting distances (repeatedly checked against known distances back home) into two categories: close (<50 m) and farther away (51–500 m), as coloration that warns a predator not to attack is likely to operate close up (Tullberg, Merilaita, and Wiklund 2005). I compare species of approximately equal weight to zebras rather than species with a similar group size simply because stalking predators and possibly even coursers are likely to have selected a single target animal at less than 50 m away. The proportions of animal sightings that I made close up (<50 m) were relatively greater for roan antelope (46.4%), hartebeest (58.5%), and waterbuck (51.5%) than for zebras (33.3%) but not for eland (43.1%) [1]. Only compared to topi (19.8%) were zebras more likely to be seen within 50 m of the car [1]. These sightings data give little support to the idea that white and black striping is particularly conspicuous, and by extension that it might warn nearby predators not to launch an attack during the day (Figure 2.1).

In section 2.1.b, I also described methods used to observe pelts disappear in the evenings and reappear in the mornings, and how pelts were observed at night. Results of these experiments showed that one zebra pelt (z2) disappeared later than the wildebeest pelt in the evenings in both wet and dry seasons, whereas the other zebra pelt (z1) disappeared after the wildebeest pelt in the wet season. In the mornings both skins appeared before the wildebeest pelt came into view. One zebra pelt disappeared later around dusk and appeared earlier in the mornings than the impala pelt when viewed through binoculars (Table 2.5). In short, zebra pelts appear conspicuous to humans in comparison to a dark gray pelage and light brown pelage, but the effect is more marked in the former case.

At night, observers found it difficult to see the pelts of these three species at a distance of 20–26 m (Table 2.6), especially when the moon was waxing. The number of nights on which observers definitely saw or believed they saw only the zebra pelt was far greater than the number of nights for the other two pelts (Table 2.7). In summary, close up to human observers, the zebra pelt was relatively easier to see than the other pelts, especially around the time of the full moon. All these quantitative data confirm that zebras are conspicuous to humans.

Data on the abilities of lions and spotted hyenas to resolve stripes presented in Tables 2.11 and 2.12 bear on aposematism. They show that under mesopic and scotopic conditions, when predators normally hunt, stripes can be resolved only at short distances. For a lion viewing the widest stripes at dusk, distances lie between 25 and 56 m or less, depending on the zebra species, and on a moonless night between 6 and 14 m or less. For a hunt to be successful, lions must break cover at <30 m (Schaller 1972), so stripes could possibly be an effective signal between 30 and 56 m at dusk, although they could not help a Grevy's zebra (Table 2.12) or any zebra species on a moonless night.

For a spotted hyena viewing the widest stripes at dusk, resolution distances are 15–34 m or less, but 4–9 m at night. The distance at which spotted hyenas begin to run after predators is very variable, ranging from 5 or 10 m up to 100 m, so stripes could theoretically act as a deterrent in those hunts that start close up, where hyenas can resolve stripes. That zebras seem to tolerate hyenas' close approach (Kruuk 1972) lends indirect support to this idea.

3.2.b Noisy behavior

Aposematic species are sometimes noisy, perhaps to enhance conspicuousness, so I wanted to look into this. I therefore recorded the number of common sounds that I could hear zebra herds make over set periods of time when I sat next to them in a vehicle. I compared this to sounds emanating from herds of impala and groups of topi (which of course are not striped) in the same area. Observations were conducted for normally one hour before 09.00 hrs or after 17.30 hrs to avoid the heat of the day, when animals rest more and make less noise; these observation periods coincide with times when herbivores can still fall prey to large predators. Specifically, I drove until I encountered a group, then I turned off the engine and sat for 15 min while the herd habituated to my presence. Herds were usually feeding and moving slowly, so I sometimes had to change position in order to keep them within sound and sight. Since group composition could alter over the observation period, I regularly counted the number of animals within sight (and sound) of me and took an average of the number of animals present over an hour. For zebra (17 h total, $\bar{X}$ herd size = 69.0), I recorded the number of snorts, neighs, bout of whinnying, and yips; for impala (18.5 h, $\bar{X}$ = 38.6), the number of snorts, grunts, clashes of horns, and trumpeting; for topi (9.8 h, $\bar{X}$ = 46.3), the number of snorts and clashes of horns. In each case I totaled these sounds and then divided by the mean number of animals in the group and by time spent watching to generate the number of sounds/capita/hour for the three species.

Overall, an individual zebra made some sort of sound nearly once per hour ($\bar{X}$ = 0.86/hr, SD = 1.90). Zebras were far more noisy than either topi ($\bar{X}$ = 0.04/hr, SD = 0.14) or impala ($\bar{X}$ = 0.09/hr, SD = 0.10) [2]. Put another way, topi were over 20 times and impala were 10 times quieter than zebras. Neighing was the most common zebra sound (N = 17 h, $\bar{X}$ = 0.68/hr, max = 7.49), although zebras also snorted ($\bar{X}$ = 0.09/hr, max = 0.57) and gave loud bouts of whinnying ($\bar{X}$ = 0.08/hr, max = 0.43) and occasionally yipped ($\bar{X}$ = 0/hr, max = 0.05). Compared to two sympatric social species, then, zebras made considerably more noise, although not specifically in response to predators.

3.3 Defense component of aposematism

3.3.a Response to predators

Zebra antipredator behavior differs according to predator species. Kruuk (1972) reported that plains zebra maintain an average distance of 40–50 m from a lion, 20–30 m from a cheetah or wild dogs, and 10 m from a spotted hyena if these predators are quietly walking across the plain. Zebras usually stand and stare at a passing carnivore, sometimes approaching a few steps (Plate 3.1). Although plains zebra will allow lions to approach to within 40 m, they will retreat at 100 m if there are foals in the group. If zebras are pursued, they flee from lions and do not usually defend the harem or foals as they do against other predators, although they can injure a lion by kicking it once contact has been made (Schaller 1972). Lions have reportedly been gored by sable antelope, roan antelope, greater kudu, and buffalo (Pienaar 1969), and killed by buffalo (Mangani 1962; Schaller 1972), but deaths or injuries due to zebras appear rare or else are not commonly documented. Regarding plains zebra, the species that has been studied most, antipredator behavior is more extensive toward midsize African carnivores than lions (Table 3.2). In general, however, once a zebra is grabbed by any predator, it shows remarkably little aggression (Kruuk 1972; Schaller 1972), perhaps because kicking is relatively powerless.

To analyze the role that stripes might serve in antipredator contexts, Hannah Walker analyzed 13 videos of a lion or members of a pride of lions hunting adult or subadult plains zebras, 10 of which ended in capture. Videos were taken from the World Wide Web because observing predation events in the wild is extremely time consuming. Of the 27 lion-zebra contacts, zebras responded by kicking the lion with both hindlegs in five instances (7, 1, 1, 3, and 8 bucks), which made contact with the lions' chests or upper ventrum 17 times, lower belly once, while two failed to make contact; a single hindleg kick was delivered to a lion's chest too. Zebras never tried to bite lions in these videos. In the 27 contacts, zebras reared up on their hindlegs on only two occasions (4 and 4 times) and on their forelegs on one occasion (3 times); in the former two instances, they successfully dislodged the lion. Taken together, these records support observations that zebras occasionally defend themselves from lions but that they are not particularly effective (see Schaller 1972). Nonetheless, it should be noted that web videos are subject to unknown bias. For instance, it could be that lions giving up hunts early are deemed of insufficient interest to post, although this seems unlikely as there are many videos of zebras simply running off by themselves.

Spotted hyenas are smaller than lions and are less able to tackle adult zebras, although they will hunt foals. A stallion will defend his family and a mare will defend her foal against spotted hyenas if they are hunting; only one zebra from a family will be involved at any one time, however. Specifically, when the harem starts to flee from hyenas, the stallion will hang back and try to drive the hyenas away. Sometimes he will run as much as 100 m after a hyena with his head near the ground, teeth bared, and ears flat. A zebra tries to bite a hyena with lowered head and to kick it with its foreleg, more rarely with its hindlegs (Kruuk 1972). Cullen (1969) reported a zebra killing a hyena with her forefeet when defending her foal, and a stallion biting a hyena on its back and casting it aside.

Wild dogs also hunt adult zebras, although less frequently than do spotted hyenas. Many of the zebra antipredator defenses are the same: chasing and attempting to bite, kicking forward and backward, stopping and facing the predator with foals standing behind (Schaller 1972; Creel and Creel 2002). Leopards, cheetahs, and brown hyenas are all smaller than lions or spotted hyenas, and they attack adult zebras infrequently and in relatively low numbers for zebras' relative abundance (see Chapter 4). In one spotted hyena, one wild dog, two leopard, and two cheetah videos of these predators hunting zebras (a very small sample), no carnivore was seen to make contact with the prey. In these videos, zebras chased hyenas with their muzzle low to the ground, offered a single kick to the hyena pack (which missed), chased the dogs for 1 sec, and made a 4 sec long chase of the cheetahs. In conclusion, if zebras are aposematic, the warning signal must surely be directed toward spotted hyenas and wild dogs or possibly smaller predators, leopards, cheetahs, and brown hyenas rather than toward lions.

To examine zebras' responses to different predators indirectly, I played predator sounds to plains zebras in the field and recorded their subsequent vigilance; zebras show heightened vigilance in the presence of large predators (Creel, Schuette, and Christianson 2014). Given the observations of others noted above, I predicted that zebras would show a marked reaction to a lion growl but would be more relaxed on hearing a spotted hyena whoop or leopard growl. In Katavi I surreptitiously placed a FoxPro High Performance Game Call speaker close to different groups of zebras that were feeding, or were drinking at a waterhole, and then retreated to the car or moved to a hiding place. I used a range finder (Bushnell Yardage Pro-Laser Rangefinder Elite 1500) to take distance readings from me to the zebras and me to the speaker and the approximate angle between them, allowing me to calculate roughly the distance between the zebras and the speaker. Then I randomly chose one out of five MP3 recordings and broadcast it through the FoxPro speaker, which was remotely controlled from the vehicle. These calls were of a lioness growling, a leopard rasping, a spotted hyena whooping; and two control species, a hippopotamus contact call and a fish eagle call, both very common in the area. These recordings had been normalized to the same peak amplitude in Raven Pro (Cornell, version 1.4) using the amplify function (set to 20,000 kU). These MP3 recordings were then exported in 16 bit wave files using Raven and then converted to an MP3 in Audacity 1.3 using the LAME encoder. These procedures simply ensured that all five calls were similar in amplitude. Calls were played briefly for a set period of time ($\bar{X}$ = 8.4 sec, SE = 0.3, range 2–20 sec, N = 108 times in total), with the amplitude of the speaker and remote control always both set to maximum.

I recorded the number of zebras in the group and the number that looked at the speaker when the call was played, and hence the proportion of group members that became vigilant on hearing the sound. I also recorded the longest number of seconds that any individual in the group

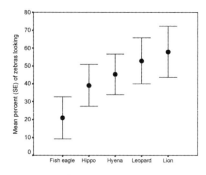 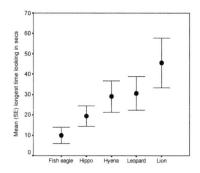

Figure 3.1. (*Left*) Mean (and SE) percentage of individuals in zebra groups that looked at the speaker following a playback call, and (*Right*) mean (and SE) longest time in seconds that a group member looked directly at the speaker. Ns refer to trials respectively involving percentage looking and the longest time an individual was vigilant: fish eagle 23, 24; hippopotamus 19, 21; spotted hyena 23, 25; leopard 22, 22; lion 21, 22.

looked at the speaker. Naively, I imagined that if zebra coloration is a deterrent to certain predator species but not others, zebras would pay little attention to the leopard and spotted hyena, predators that have possibly learned zebras are aposematic prey, and almost no attention to the control sounds, but much attention to the lion call, as lions are their chief predator.

Zebras showed different responses to the five different playback calls as measured by the percentage of group members that were vigilant and the maximum length of time that a group member looked at the speaker (Figure 3.1). For both measures there was a rank ordering of intensity of response from fish eagle to hippopotamus (the control sounds), to spotted hyena to leopard to lion. In both statistical models [3, 4], call type was an important factor, but distance from the speaker and to some extent call duration also affected the proportion of animals that were vigilant. Interestingly, the time that the most vigilant zebra spent looking at the speaker when it heard a spotted hyena whoop was shorter than when it heard a lion growl [5], and somewhat shorter when it heard a leopard growl than a lion growl [6]; responses to hyena and leopard did not differ [7]. Therefore, zebras seemed to find the presence of a hyena or leopard of less concern than a lion as judged by this measure of attentiveness. This suggests that plains zebras perceive predator species as differentially threatening, which might be linked to differences in prey-predator relative body sizes or, more relevant to the aposematism idea, to zebras' ability to defend themselves against smaller predators, perhaps advertised by conspicuous stripes.

Biting

Teeth can be effective weapons against predator attack and during intra-specific fights. Compared to ruminants of similar size and diet, equids have larger masseter muscles and mandibular masseter insertion area (Turnbull 1970) and have more robust jaws (Fletcher, Janis, and Rayfield 2009). Total bite force in horses is known to be almost double that in cattle (Hongo and Akimoto 2003). Stallions and mares both have canines that are not found in artiodactyls. Additionally, equids have both upper and lower incisors, whereas artiodactyls have no upper incisors. Last, because equids are grazers, they have relatively broader muzzles than browsers, although thinner than grazing artiodactyls (Janis and Ehrhardt 1988). All these points suggest that a bite from an equid can inflict substantially more damage than from an artiodactyl. Recall that children are told to be careful feeding domestic horses for fear of having their fingers bitten.

Hannah Walker took measurements of teeth heights and muzzle widths from museum skulls. Specifically, (i) she measured the heights of incisors on the lingual side starting at the center of the tooth at the root to the center of the tooth at the tip; (ii) behind the incisors she measured the shortest distance from the outer root corner of the outer incisor to the opposite outer incisor to give the diameter of the incisor span; and (iii) she measured length of canine teeth. Assuming that a zebra can bite off a half cylindrical piece of flesh from a predator equal to the height of incisors and the diameter of their curvature, I multiplied incisor height by πr^2 (where r is the incisor span/2). I did this for both upper and lower incisors in equids but only for the lower incisors in sympatric artiodactyls, as they lack upper incisor teeth. These are crude first approximations of the danger of being bitten. Table 3.4 shows that most sympatric herbivores can remove only a quarter as much flesh as can a zebra; only African buffalo are able to dislodge a chunk of flesh equivalent to that bitten off by a zebra. Since grazing Grevy's zebra have incisor sharpness similar to that of ruminants (Popowics and Fortelius 1997), these rough calculations suggest that being bitten by a zebra is potentially far more damaging than being bitten by a similar-size ruminant.

Turning to comparisons simply within equids, the volume of flesh that a zebra can remove is not, however, substantially different from unstriped equid species; indeed, Przewalski's horse has a far greater bite volume than any other equid (Table 3.4). Furthermore, there are no differences in canine lengths between striped and unstriped adult and subadult equids as measured from root of the longest canine to its apex (Table 3.5). To summa-

Table 3.4. Incisor lengths and calculated biting capabilities (see text) of adult and subadult equids combined and artiodactyls sympatric with zebras (Ns in parentheses).

	Incisor heights in mm		
	Upper	Lower	Calculated bite volume (cm³)
Plains zebra	18.9 (14)	18.1 (14)	50.2
Mountain zebra	18.4 (6)	20.0 (5)	44.5
Grevy's zebra	19.4 (12)	18.1 (14)	53.6
Blue wildebeest	—	11.4 (6)	12.1
Roan antelope	—	14.9 (6)	12.3
Oryx	—	12.8 (9)	10.3
Hartebeest	—	11.3 (9)	76.1
Waterbuck	—	11.0 (10)	53.5
Topi	—	10.4 (7)	69.9
Buffalo	—	19.7 (6)	44.1
African wild ass	17.3 (4)	17.1 (4)	35.9
Asiatic wild ass	17.5 (8)	18.8 (7)	44.8
Przewalski's horse	23.9 (4)	25.8 (4)	86.5

Source: H. Walker, unpublished data.

Table 3.5. Mean upper and lower canine lengths in mm of male and female adult and subadult equids combined (Ns in parentheses).

	Males		Females	
	Upper	Lower	Upper	Lower
Plains zebra	18.7 (5)	19.6 (6)	6.6 (6)	8.3 (5)
Mountain zebra	17.1 (2)	17.8 (2)	—	—
Grevy's zebra	14.8 (8)	19.5 (6)	9.3 (4)	12.4 (3)
African wild ass	13.6 (2)	15.7 (2)	—	—
Asiatic wild ass	15.2 (2)	11.7 (2)	11.8 (1)	7.6 (2)
Przewalski's horse	19.3 (3)	19.9 (2)	5.9 (1)	—

Source: H. Walker, unpublished data.

rize these data, zebra mouth morphology gives them a formidable ability to bite a tormentor. Despite this, written accounts and the videos that we examined indicate that biting is rare.

Wounding

If zebras are aposematically colored, one might expect them to be attacked infrequently. To determine whether zebras escape attention of predators through warning coloration, I compared the number and extent of wounds that I saw on live adult and subadult zebras, giraffe, buffalo, impala, waterbuck, topi, reedbuck, and warthog over the course of three years in Katavi. I chose a wide three-year time window in case individual predators—say, a pride of lions—were specializing on a particular species in a given year. Individual animals were scored as being healthy or wounded on either their left or their right side, as I could normally only see one aspect in the course of my activities in the park, and I did not want to disturb the animals by driving around them.

A large proportion of adult and subadult zebras carried both fresh wounds and old scars. I saw fresh gashes up to an estimated 60 cm in length, every so often in parallel, suggesting they had been sculpted by claws, and were at times deep enough for the red muscle to be seen below (Plate 3.2). Old scars were sometimes long and straight or slightly curved, visible because the black or white stripes had not realigned (Plate 3.3). Often tails were severed, suggesting they had been bitten off, as described for black rhinoceros (Plotz and Linklater 2009). Generally, wounds were more often found on the rump, haunches, and tail than elsewhere on the body (Table 3.6) and were more prevalent in these regions in zebras than in all the other species combined [8]. Zebra wounds were certainly more severe than in other ungulates (Appendix 2), and I found it difficult to believe that some individuals could even walk given the depth and length of wounds that they must have received in the previous few hours. I suspected that most wounds had been inflicted by lions because they were high up on the body, which necessitated climbing up the animal, and because I saw lions frequently in the area. The location of wounds on the zebras' bodies fitted well with Schaller's (1972, p. 264) description: "The lion lunges and places both forepaws on the zebra's rump, which buckles under the impact."

Rates of wounding were far greater in zebras than similar-size herbivores. Of the 850 right or left sides of zebras that I observed, 69 sides (8.1%) showed signs of wounding. This proportion is much greater than species of approximately similar body weight, topi (0/105 sides) and waterbuck (0/75 sides), and greater even than smaller heterospecifics: reedbuck (0/43 sides), impala (12/886 sides, 1.4%), and warthog (1/45 sides, 2.2%). The proportion was similar to that of other, much heavier, species—namely,

Table 3.6. Distribution of wounds on four species of herbivore in Katavi National Park.

Species	Rump/haunch	Tail	Flank	Other area
Zebra	47	10	5	6
Giraffe	10	6	6	4
Buffalo	1	6	2	2
Impala	2	0	5	2

giraffe (24/339 sides, 7.1%) and buffalo (10/109 sides, 9.2%). Note, however, that the buffalo figure is likely to be artificially high because I usually counted old males that are often the focus of lion attacks (Schaller 1972); it was difficult to get close enough to observe the more flighty female buffalo that live in large herds. In short, in Katavi National Park, plains zebras receive more attacks than other species of similar body weight, a finding not commensurate with aposematism. The key comparison would be of striped and unstriped equids, which cannot unfortunately be achieved because of conservation and ethical reasons.

3.4 Conclusion

The principal reason for considering zebras as aposematic hinges on their repeated, contrasting, and conspicuous black and white stripes possibly being a warning signal. Information presented in this chapter gives little support for this idea. During the day, zebras are not markedly more conspicuous (to humans) than other herbivores close up, although they are notably more noisy. For humans, their pelts are easier to see than gray or light brown pelts in the early mornings, evenings, and at night, but for lions and spotted hyenas my work with Amanda Melin, Don Kline, and Chihiro Hiramatsu indicates that stripes are only resolved at close distances, distances at which these predators have often already committed themselves to a chase, thus making aposematism ineffective.

Additional strands of data presented here lend little support to the deterrent component of aposematism. Although zebras show graded responses to predator species, being relatively indifferent to spotted hyenas but wary of lions, they are regularly attacked and wounded by lions, suggesting that striping does not deter attacks per se. Moreover, they have no weaponry such as horns that can be used in defense (Stankowich and Caro 2009), and they resort to biting predators infrequently, although their hooves are dangerous. Certainly there are behavioral characteristics that comply with being aposematic, including aggregation, conspicuous behavior, and absence of hiding behavior, but these are found in many nonstriped artiodactyls too. Contrary to other aposematic mammals, zebras are not sluggish and can run fast and for long distances. In sum, zebras lack the necessary obvious defenses to make them aposematic.

In nature, the efficacy of aposematism depends on numerous factors, including predator hunting strategies (Endler and Mappes 2004), the predator's state of hunger (Mappes, Marples, and Endler 2005), relative size of predator and prey (Hagman and Forsman 2003), prey toxicity (Darst, Cummings, and Cannatella 2006), and prey abundance (Speed and Ruxton 2007), so we should not expect zebras' coloration to deter all predatory attacks (Beddard 1892). The fact that spotted hyenas, brown hyenas, wild dogs, leopards, and cheetahs occasionally attempt to pursue, catch, and kill zebras in some ecosystems, although not particularly successfully, is not therefore an impediment to the aposematism hypothesis (see Chapter 4), but at this stage it might be most cautious to simply remember that many conspicuous animals are not necessarily aposematic (Ruxton, Sherratt, and Speed 2004).

Statistical tests

[1] Number of sightings of zebras seen close, within 50 m of the transect line, and farther away, 51–500 m of the transect line, compared with eland ($\chi^2 = 1.934$, df = 1, NS); roan antelope ($\chi^2 = 4.063$, df = 1, p < 0.05); hartebeest ($\chi^2 = 9.455$, df = 1, p < 0.001); waterbuck ($\chi^2 = 17.453$, df = 1, p < 0.001); and topi ($\chi^2 = 9.050$, df = 1, p < 0.01).

[2] Comparison of number of sounds/individual/hour made by zebra, topi, and impala, Kruskal-Wallis test $\chi^2 = 11.680$, df = 2, p = 0.003. Mann-Whitney U tests, zebra vs. topi, U = 31.0, p = 0.002; zebra vs. impala, U = 92.5, p = 0.027; topi vs. impala, U = 61.0, p = 0.045.

[3] Percentage of group that looked at the speaker following a playback as a function of call duration (F = 3.350, df = 1, 99, p = 0.063), distance from the speaker (F = 5.144, df = 1, 99, p = 0.025), and call type (F = 5.791, df = 4, 99, p < 0.0001) using a univariate analysis of variance.

[4] Longest time spent looking at the speaker as a function of call duration (F = 0.029, df = 1, 101, NS), distance from the speaker (F = 1.258, df = 1, 101, NS), and call type (F = 9.952, df = 4, 101, p < 0.0001) using a univariate analysis of variance.

[5] Comparing longest time spent looking at the speaker following a spotted hyena call or a lion call as a function of call duration (F = 0.003, df = 1, 40, NS), distance from the speaker (F = 1.529, df = 1, 40, NS), and call type (F = 4.206, df = 1, 40, p = 0.047) using a univariate analysis of variance.

[6] Comparing longest time spent looking at the speaker following a leopard call or a lion call as a function of call duration (F = 0.016, df = 1, 38, NS), distance from the speaker (F = 0.837, df = 1, 38, NS), and call type (F = 3.083, df = 1, 38, p = 0.087) using a univariate analysis of variance.

[7] Comparing longest time spent looking at the speaker following a spotted hyena call or a leopard call as a function of call duration (F = 0.209, df = 1, 42, NS), distance from the speaker (F = 2.658, df = 1, 42, NS), and call type (F = 0, df = 1, 42, NS) using a univariate analysis of variance.

[8] Wounds on the rump, haunch, and tail vs. other parts of the body were greater in number for zebras than other species—i.e., giraffe, buffalo, and impala combined, $\chi^2 = 15.570$, df = 1, p < 0.001.

Figure 4.0. Stripes presented by many individual zebras fleeing together are thought to confuse the predator in a number of different ways. (Drawing by Sheila Girling.)

FOUR

Predation and confusion

4.1 Confusion

The idea that zebra stripes confuse predators is perhaps the most popular reason that the public holds for why zebras are striped (Plate 4.1). A predator can become confused by the behavior of its prey due to erratic movement or because the prey moves in a way that exploits weaknesses in the predator's visual system (Stevens 2013). More formally, the confusion effect in biology is defined as the reduction of the attack-to-kill ratio in predators owing to an inability to single out and attack an individual when prey are aggregated (Krause and Ruxton 2002). Following multiple prey targets may result in increased visual or cerebral processing by a predator that decreases its search efficiency (Tosh, Krause, and Ruxton 2009). This could even lead to a reduction in attack rate because the predator may incur costs trying to overcome confusion, lowering attention paid to personal predation risk, for instance (Milinski 1984). Second, if an attack is subsequently launched, the probability of success may be reduced (Neill and Cullen 1974; Cresswell 1994). These two aspects of predator confusion can operate within the same predator-prey system—for example, in three-spined sticklebacks feeding on *Daphnia* (Ioannou et al. 2007)—and both may be exacerbated by larger prey group size (Krause and Ruxton 2002) and, crucially in zebras, when many individuals have a similar appearance (Krakauer 1995).

The means by which zebra stripes might confuse predators actually come in six guises (Morris 1990). The first two turn on the idea that stripes of one individual are difficult to distinguish from stripes on a second individual in situations in which there is no gap of light between animals. Under such circumstances it might be difficult for a predator to ascertain the number of zebras in a herd and therefore whether it is worth attempting to hunt. Second, it could be difficult for a predator to discern the true outline of a single individual when a group flees (Hall et al. 2013). A third idea is that stripes make it difficult to follow a single individual during a chase (Eltringham 1979; Kingdon 1984). A fourth rather different conjecture is that black and white stripes dazzle a predator when viewed close up, making it difficult to concentrate on the quarry and preventing it from being followed or making it uncomfortable to look at (Morris 1990). This rather vaguely formulated idea refers to moving but possibly stationary prey as well. A fifth idea is that zebra stripes moving across the retina of an observer generate motion dazzle, where their markings make speed and trajectory difficult to estimate (Stevens and Merilaita 2009a). It has received much recent attention stemming from a series of experiments using human subjects trying to target moving striped objects on a computer screen. Last, it is thought that striping may cause an attacker to misjudge the size of prey and either deter it from attacking because of its relatively large size compared to the predator, or cause the attacker to miscalculate its final leap and miss the target (Cott 1940). I now go through these forms of confusion one by one.

4.2 Miscounting numbers of prey individuals

Generally, predators choose to hunt small to medium-size groups of prey over larger ones and so must be able to assess group size in some way, although we know little about their actual computational abilities (but see McComb, Packer, and Pusey 1994). We do know that it is more difficult for people to count accurately larger numbers of objects, and similarly to count larger groups of animals (Norton-Griffiths 1978), so this might also apply to large predators. Striping might make counting additionally difficult if it prevented an observer from telling individuals apart, particularly in large groups. To begin to explore this issue, I investigated whether human observers found it more difficult to count zebras than other species. So I revisited observations that I and two others made on 10 sympatric species of large herbivores—elephant, buffalo, giraffe, hippopotamus, zebra, topi, waterbuck, warthog, impala, and reedbuck—usually grazing at distances of 300 m or farther from us during periods of low light intensity in late evenings and early mornings (section 2.1.b). As described previously, observers sat on the side of an open floodplain and used 8 × 30 binoculars to record numbers of individuals of each species every 5 min; different species were present on different days, but zebras were always present. Observers made records in notebooks privately. Observations were conducted for a total of 26 mornings and 28 evenings and started before dawn or before dusk. Since animals were usually observed together in one herd of conspecifics that remained the same size during these observation periods, I used variability across our repeated counts as a crude measure of the difficulty that observers found in counting striped and nonstriped species under low light conditions when hunting occurs. If zebras confused the eye when they were close together (in its simplest form reducing two adjacent individuals to one), one would expect greater variation in zebra numbers over repeated counts during an observation session.

I separated the data into four group size categories as estimated by the mean of the counts during an observation session (small 1–10, medium 11–50, large 51–100, very large 101–200 individuals) because this type of confusion effect would be expected to be most marked in larger-size groups. This is because more individuals would be likely to overlap in the sense of showing no light between them from an observer's point of view. I then compared the standard errors in the three observers' counts during each evening watch and each morning watch for every species of herbivore. There was no effect of species but a large effect of group size, with counts of larger groups being far more variable irrespective of species [1]. Therefore, striping seems to have no effect on human observers' abilities to count zebras accurately. Personally, I did not find this surprising because at far distances I could not see stripes, but I could identify a herbivore as being a zebra by its trapezoid shape, the characteristic shape of its thick neck and head, and the frequent swishing of its tail (see Chapter 5).

In passing, stripes of individual zebras eliding with one another are more likely to reduce the apparent number of individuals in a group than to increase it, which would make the herd marginally more attractive to lions, which prefer to hunt smaller groups of prey (Scheel 1993), and therefore be disadvantageous to zebras.

4.3 Striping obscuring outlines of fleeing prey

In circumstances where a single predator or a group of predators target a single prey item, as is common among spotted hyenas and lions hunting alone or in groups, the predator will need to detect the outline of an individual prey item before it decides which animal to pursue. For stalkers this might occur while the prey is stationary or is beginning to run away, while for coursers this might occur while the group is fleeing. In either case, several animals in a line could generate a continuous pattern of vertical stripes, making such discrimination difficult. I therefore made the following simple predictions: if striping serves to obfuscate the outline of an individual, then zebras should be bunched when stationary or during flights, should run off in a more or less continuous line of animals, and should leave few gaps of light between them. These are necessary conditions for striping being a means to confuse predators that are hunting.

To test these predictions, I drove near harems and herds of zebras, not lone individuals, in my vehicle and stopped the car some distance away from them. I then took a distance reading with a range finder, counted the number of individuals in the group, and estimated, from where I sat, the maximum number of body lengths separating any two neighboring zebras in the group. Next, I got out of the far door of the vehicle and slowly walked around it and from there set out walking in a determined way directly toward the group, counting off the number of paces to myself as I did so. This allowed me to record flight initiation distances by subtracting paces from the original range finder reading. As soon as the group bolted, I stopped and recorded the number of animals fleeing and whether group members moved toward one another (bunched), moved apart rapidly (exploded), or maintained approximately the same distance from one another as before the flight (no change). With a stopwatch, I timed the flight to the nearest second and subjectively estimated flight length in meters by eye. During the course of the entire flight, I scored whether or not zebras presented a more or less continuous line of stripes to me, and in those flights where they did, I estimated the shortest nearest neighbor distance in number of body lengths. This gave me an approximate measure of group cohesion throughout the flight. In addition, I estimated the maximum number of body lengths (0 body lengths, 1–2 body lengths, and ≥3 body lengths) that separated any two adjacent individuals at 5 sec into their flight to give an approximate measure of dispersion at a snapshot in time after fleeing. If animals were trying to confuse a predator, one would expect no gaps between individuals after 5 sec into the flight.

First, I should summarize the basic data. I encountered and approached zebra groups in this way 104 times in total. Zebras' average group size was 8.8 (SE = 0.7; range 2–47 animals), and the median group size was 6, indicating that I usually approached harems. I made most of my approaches in grassland (68.3%) or in light woodland (18.3%), but a minority were in wooded grassland (6.7%), bushed grassland (4.8%), or bush (1.9%). Zebra groups always fled from me, and in most situations all the group members ran off together ($\bar{X}$ = 93.3% of group members, SE = 1.9). Groups fled at an average of 75.1 m (SE = 3.9, N = 104 flights) from me (i.e., flight initiation distance), and ran for an average of 13.3 sec (SE = 1.1, N = 102 flights). I was concerned that my personal estimation of flight length ($\bar{X}$ = 52.2 m, SE = 5.5, N = 89 flights) was inaccurate, so I divided my estimated flight length by recorded flight time to generate estimates of flight speeds, and reclassified the resulting speeds into broad categories in some analyses: slow (1–3 m/sec), medium (3.1–5.0 m/sec), and fast flights (5.1–10.0 m/sec). When I did this, I found that 31.0% of zebra flights occurred at a slow rate, 58.6% at a medium rate, and 10.3% at a fast rate.

4.3.a Lines of stripes shown to humans

Prior to flights, I usually encountered stationary groups with neighbors standing fairly close to one another (<1 body length apart on 30.2% of occasions) or between 1 and 3 body lengths apart on 46.5% of occasions (N = 86 groups). Zebras did not form tightly knit groups during flights and did not crisscross one another across my field of view. Groups bunched together on only 31.7% of 104 flights, exploded on just 6.7%, but mainly showed no marked change in group cohesion (61.5%). Therefore, there was the potential of presenting a continuous line of stripes to me in only about one-third of their flights. Yet, on closer inspection, a group never actually presented anything resembling a continuous line of stripes to me in 53.5% of 99 flights. On the remaining 46.5% of occasions where groups were reasonably cohesive (amounting to 45 flights, as 1 had to be dropped), I saw a continuous line of stripes (that is, 0 body lengths between all neighbors) on a mere 15.6% of those 45 flights. In the remainder I could discern gaps between individuals either of 1 body length (44.4%), 2 body lengths (22.2%), or 3–4 body lengths (17.8%). Overall, then, I saw a continuous line of stripes with no gaps between individuals in just 7 out of 99 flights from me.

Examining my second measure of cohesion, at 5 sec into the flight, I recorded no gaps between the two adjacent individuals in 23.0% out of 61 flights in which I took this measure, gaps of 1–2 body lengths between two neighbors on 26.2% of flights, and gaps of 3 or more body lengths between the closest two adjacent individuals on 50.8% of flights. In summary, zebras stayed close together when watching me approach but then usually fled as a loosely cohesive group but with gaps between individuals. These observations (based on reactions to me) suggest that predators will encounter individuals reasonably close together, but once the predator breaks cover, it will be easy to single out an individual during its pursuit. My two sorts of data speak against the second type of confusion effect in which it might be difficult to pick out a prey individual from the herd (see Ruxton, Jackson, and Tosh 2007).

4.3.b Lines of stripes in dangerous situations

If striping is a means by which zebras confuse predators by making it difficult to single out one individual from the herd, one might expect zebras to make a greater effort to obscure their outlines in more dangerous situations. In other words, zebras might flee in a more cohesive fashion when I was close and flight initiation distances were short. There was, however, no effect of flight initiation distance on whether group members bunched during flight [2], on whether I could see a more or less continuous line of animals [3], on whether there were no gaps between the closest two individuals if group members paraded a more or less continuous line of stripes [4], or on gap width at 5 sec into the flight [5]. Rather, I was marginally more likely to see a continuous line of stripes when fewer zebras fled and when they ran off at slower speeds [4]; while at 5 sec into the flight, I was more likely to see no gaps between two neighbors during slower flights [5]. Put another way, gaps were more likely to be seen during faster flights from me when zebras presumably judged me as being more dangerous. This runs contrary to the idea that under risky situations zebras use behavioral means to enhance a putative confusion effect of stripes during flight. There remains, however, the possibility that zebras react differently to a person on foot than to a natural predator, so we examined video footage of herd striping patterns while being hunted by lions.

Hannah Walker watched a total of 25 videos of plains zebras fleeing and another 25 of plains zebras fleeing from one or more lions—the first set to obtain an independent data set from my experiments, the second to observe zebras under naturally dangerous circumstances. Real-time videos were separated according to whether side views or hind views of zebras could be seen. Measurements were taken every second using a detailed timer on the video or else recording to the nearest second by stopwatch. For videos showing zebras' lateral surfaces, the number of times any two adjacent zebras were separated by more than one body length, approximately one body length, less than one body length, or by no gap at all were scored each time the video was halted. For hind views, the number of times every two adjacent individuals were separated by more than one rump width, by approximately one rump width, by less than one rump width, or overlapped with another zebra were scored. On those occasions where there were no gaps between two animals in a given frame, additional information was taken to understand the nature of alignment. For side view videos, a note was made of (i) the number of instances where black vertical neck or flank stripes formed a continuous pattern across overlapping animals, or (ii) where horizontal rump stripes were aligned with those of an adjacent zebra, or (iii) where oblique black stripes on the rump overlapped with the oblique black stripes on a neighbor's rump, or (iv) where black stripes were not aligned, or (v) where overlap was indeterminable or the frame was blurred. The same procedure was carried out for hind view videos, recording whether (i) horizontal stripes were aligned or not when adjacent zebras' rumps overlapped with each other, or (ii) where this could not be determined or the frame was blurred.

Similar to flights from the author, the proportion of flights in which gaps could be discerned between adjacent individuals was high. On almost two-thirds of occasions, neighbors' bodies did not overlap when observed from the side, and on three-quarters of occasions when fleeing directly away from the camera whether a lion was in the video or not (Table 4.1). In those circumstances where adjacent individuals' side views did overlap, black stripes were aligned on an average of 68.0% of the frames of the 16 sampled videos in the absence of a predator and significantly more ($\bar{X} = 88.6\%$ in frames of 14 videos) when one or more lions were taking part in a hunt [6]. In flights where stripes were aligned, it was usually the oblique stripes across the rump that formed a continuous pattern between neighbors (no lions 38.7%, lions 64.3%), rather than a horizontal (16.8%, 20.3%, respectively) or vertical alignment (12.4%, 4.0%, respectively)

Table 4.1. Mean percentages of frames in videos in which adjacent zebras were separated by different distances as measured in body lengths.

	No lions	Lions
Side views		
Number of videos	19	17
Average total number of adjacent animals scored in all frames per video	66.5	64.8
% > 1 body length	16.6	17.9
% ≈ 1 body length	8.3	10.1
% < 1 body length	37.8	33.5
% No gap	37.4	38.6
Hind views		
Number of videos	6	8
Average total number of adjacent animals scored in all frames per video	16.3	13.3
% > 1 rump width	34.8	47.4
% ≈ 1 rump width	13.4	4.9
% < 1 rump width	33.5	25.9
% No gap	18.4	21.7

Table 4.2a. Number of instances (i.e., frames where videos were stopped every 1 second) in which there was no gap between side views of adjacent zebras and in which their stripes were aligned at different angles.

Video	NF	H	V	O	NA	U	B
Lions present							
A	29	0	0	11	8	10	0
B	100	13	0	15	4	15	53
C	5	1	0	1	0	1	2
D	6	2	0	1	1	1	1
E	6	2	0	2	2	0	0
F	15	2	0	3	0	0	10
G	1	0	0	1	0	0	0
H	61	4	0	30	16	11	0
I	7	1	0	2	0	0	4
J	7	0	0	6	0	0	1
K	1	0	0	0	0	0	1
L	57	3	0	11	0	9	34
M	6	0	1	3	0	0	2
N	42	2	4	17	4	15	0
Lions absent							
O	7	0	0	3	4	0	0
P	25	6	0	7	1	4	7
Q	13	2	4	4	1	1	1
R	10	2	0	0	3	1	4
S	20	0	1	9	4	5	1
T	68	11	0	3	5	11	38
U	7	0	2	1	2	2	0
V	5	1	1	0	2	1	0
W	17	2	0	6	2	7	0
X	8	0	0	3	5	0	0
Y	26	6	1	4	4	11	0
Z	26	0	2	12	5	7	0
AA	55	7	3	24	16	5	0
BB	68	5	0	28	8	11	16
CC	4	0	0	2	1	0	1
DD	29	0	0	11	8	10	0

Source: H. Walker, unpublished data.
Note: Column headings are as follows: NF number of frames; H aligned horizontally; V aligned vertically; O aligned obliquely; NA not aligned; U cannot determine; B blurred.

(Table 4.2a), and oblique stripe alignment was greater in lion videos than no-lion videos [7]. For hind views where rumps overlapped, neighbors' horizontal rump stripes were aligned in an average of 82.5% of frames taken from 4 flights with no lions and 100% of frames in the 6 videos when lions were hunting [8] (Table 4.2b). In summary, there were few instances where zebras' stripes formed continuous lines in dangerous situations, again giving little support to the second type of confusion effect. If stripes were to obscure the outlines of fleeing zebras, they would likely be alignments of rump stripes viewed from the side or behind.

Table 4.2b. Number of instances (i.e., frames where videos were stopped every 1 second or 0.5 second, depending on the film speed) in which there was no gap between hind views of adjacent zebras and in which their stripes were aligned horizontally.

Video	NF	H	NA	U	B
Lions present					
AAA	1	1	0	0	0
BBB	6	3	0	1	2
CCC	3	3	0	0	0
DDD	2	2	0	0	0
EEE	6	5	0	1	0
FFF	7	6	0	1	0
Lions absent					
GGG	5	4	1	0	0
HHH	2	1	1	1	2
III	1	1	0	0	0
JJJ	1	1	0	0	2

Source: H. Walker, unpublished data.

Notes: Column headings are as follows: NF number of frames; H aligned horizontally; NA not aligned; U cannot determine; B blurred.

4.4 Striping preventing a single prey individual being followed

Unpredictable flight paths, or a high variance in turning angle, called protean behavior (Humphries and Driver 1967), have been shown to reduce capture rates in humans attacking moving prey on a computer screen, an effect that has been shown to be both related (Scott-Samuel et al. 2015) and unrelated to group size (Jones, Jackson, and Ruxton 2011)! Thus, one might argue that zebras would flee in an erratic fashion if striping hinders a predator from visually tracking an individual zebra during pursuit of a group. For instance, it is possible that zebras' greater speeds and sharper turns at night in grassland than in woodland is a response to danger posed by lions there (Fischhoff et al. 2007), although these behaviors were not recorded in the context of hunting but in risk management. I therefore examined a number of measures of protean behavior in flights of zebra groups from me, specifically whether groups exploded, ran off at a sharp angle, or zigzagged, or if fleeing individuals switched rear position in the group, the position that is closest to a predator during flight.

I used the same method of walking toward groups of zebras as described earlier, recording flight initiation distances, number of animals fleeing, flight time and length, and the pattern of flight (whether fleeing group members bunched or exploded), and estimated whether the group fled directly away from me or 10° to either side (called directly), at an angle of 10–45° from my direction of travel, or at a sharp angle of 46–90°. I also recorded whether the group zigzagged, which I defined as making a sharp movement at an estimated 45° or more from the line of travel, since this might make individuals more difficult to track. As a pursuing predator might find it easier to concentrate attention on the nearest individual to it, I recorded whether the individual at the rear of the group switched to another position in the herd during flight.

As reported earlier, explosive departures from me were relatively uncommon (6.7%). Second, zebras fled in all directions: 31.7% directly away from me, 25.0% at estimated 10–45° angles, and 43.3% at estimated 46–90° angles (N = 104 flights). Third, whatever the direction of travel, the group ran in a straight line or arc from me in the majority of flights (82.4% of 102 flights) and rarely zigzagged (17.6%). Fourth, in most flights (69.3% of 75), the rear individual did not switch position. None of these findings support protean behavior.

Protean behavior was not manifested in dangerous situations. When I was close and presumably more of a risk, zebras fled directly away from me showing their rumps, while at longer flight initiation distances, group members were more likely to flee at a sharp angle from my direction of travel [9]. Nonetheless, at shorter flight initiation distances, the likelihood of zigzagging was greater, and zigzagging was marginally more likely when flights were of longer time duration or if they were fast [10]. The probability of rear position switching was unaffected by flight initiation distance, number of zebras fleeing, flight time, or flight speed [11]. To sum up these findings, behavioral means of confusing predators were used very sparingly during flights and in only one circumstance were they associated with danger: zigzagging was more likely when flight initiation distances were shorter. The third form of confusion, that striped pelage is associated with unpredictable flight behavior, is not strongly supported.

4.5 Dazzle effect

Zebras' coats did not appear to deceive lions from making accurate contact with prey. Hannah Walker analyzed 13 videos from the web of lions attacking adult and subadult zebras, 10 of which were successful. There was not a single instance in which the lion missed its quarry while trying to contact the prey. Of course, videos posted on the web are a selected sample, but the unanimity of successful contacts speaks against stripes dazzling the lion. In some videos a lion made multiple contacts with the quarry because the zebra broke loose from it. Also, some videos showed several lions making contact with the prey. From a total of 27 contacts, then, two were made on the stomach, three on the rump, twelve on the back, one on the shoulder, and nine on the neck. In one spotted hyena video, one wild dog, two leopard, and two cheetah videos, no attempted contacts were recorded. Judging from the predators' behavior, there is no evidence that they misjudged their attack on zebras due to stripes or other factors.

4.6 Motion dazzle

The fifth form of confusion is motion dazzle. Motion dazzle refers to misjudging the speed and trajectory of a moving object. First proposed by G. Thayer (1909) and A. Thayer (1918), the idea that large blocks of color could disguise the direction and speed of an object was developed by Wilkinson and others and applied to transport vessels crossing the North Atlantic in World War I (Behrens 2002). Ships painted in this way were thought to be difficult targets for distant submarines that had to estimate the direction of the ship before firing their torpedoes in advance of the ship's path (D. Williams 2001). There are putative examples of motion dazzle in reptiles (Jackson, Ingram, and Campbell 1976) and fishes (Marshall 2000) moving rapidly, but these are somewhat speculative. Recently, attention has focused on empirical studies with humans and the implications that these might have for moving zebras. These studies provide mixed support for stripes being a form of motion dazzle, and they deal mostly with single prey, not groups with static patterns, as follows.

In a computer game involving people trying to "catch" one achromatic "prey item" at a time against leafy or grassy artificial backgrounds, an unpatterned target matching the background was most difficult to "capture" with a mouse click. Dazzle (i.e., here static black and white oblique, vertical, and horizontal stripes), horizontally striped, vertically striped, or zigzag "prey types" that moved across the screen were more difficult to catch than a uniform conspicuous target but easier than a white target (Stevens, Yule, and Ruxton 2008). In a second experiment against a leafy backdrop, prey with broad banded vertical stripes were more difficult to catch at high and low speeds than prey with thinner banded stripes. In a subsequent set of experiments using a camouflaged background, moving prey with vertically striped markings were captured less often than uniform conspicuous white targets or camouflaged targets, but they did not differ from a gray target. An additional experiment showed that these effects were more pronounced when targets were of low contrast (Stevens et al. 2011). In a third study, again using human volunteers, subjects were presented with two moving stimuli and asked to report which was moving faster across a computer screen. Compared to a plain control, subjects reported that zigzag and checkered patterns moved about 7% slower than the control in a high-contrast condition when objects were rapidly moving across the screen, but this was not the case for horizontally or vertically striped stimuli. Movement in this experiment was equivalent to an object moving at 13 km/hr (3.6 m/sec) at a distance of 10 m. At low contrasts or when movement occurred at slower speeds, there was no perceived difference between stimuli (Scott-Samuel et al. 2011).

In a fourth study, von Helversen, Schooler, and Czienskowski (2013) showed that striped objects were perceived as moving faster than unicolored objects but were hit more, particularly if they were longitudinally striped. The relationship between speed perception and hit rates is therefore not straightforward and may depend on the experimental setup. Yet another study demonstrated that striped targets were more difficult to catch than white, spotted, and camouflaged targets but no more difficult than gray or white-edged gray targets, although striped targets were attempted relatively quickly by subjects (Hughes, Troscianko, and Stevens 2014). Finally, research examined target stripe orientation and number. When single targets were presented sequentially, parallel striped targets were easiest to catch against a leafy gray background; perpendicular and obliquely striped targets were as easy to capture as gray targets. When targets were presented in groups of six, however, moving independently across the screen, they were attempted more rapidly than gray targets, and fewer capture attempts were required to catch them (Hughes, Major-Elliott, and Stevens 2015).

Some of these laboratory findings suggest that the speed of moving striped objects might be difficult to judge, but the stimulus designs in these experiments resemble only parts of a zebra (Table 4.3). For example, a lateral view of a living zebra consists of vertical, oblique, and horizontal stripes, while a distal view shows horizontal stripes that curve down near the body's outline (and zebras do not have zigzag, checkered, or box-shaped lines). Second, if stripes were to cause a predator to misjudge its leap at a zebra, it would operate very close to the predator, whereas the prey in the laboratory experiments subtend a very small angle of the human eye. In future studies, the stimuli require more realism.

How and Zanker (2013) explored the mechanism by which motion dazzle in zebras might operate. Using a motion detection algorithm to analyze motion signals generated by different areas of the zebra's body

Table 4.3. Summary of findings of motion dazzle experiments with human subjects and their relevance to the function of zebra stripes.

Principal findings	Points that are relevant to zebra striping
Stevens, Yule, and Ruxton (2008)	
Bands and zigzag markings difficult to catch. Prey more difficult to catch at faster speeds and on heterogeneous backgrounds. Broad band vertical markings more difficult to capture than thinner markings.	**Horizontal and vertical patterns resemble rump and flank markings.** *Gray markings more difficult to catch than patterned targets.*
Stevens et al. (2011)	
Vertically striped markings more difficult to catch than camouflaged or uniformly conspicuous markings. Targets more difficult when of low contrast.	**Striped moving targets caught less often than background-matching targets.** *Gray markings as effective as striped prey in reducing capture.* *Zebra markings are of high contrast.*
Scott-Samuel et al. (2011)	
Checked or zigzag markings perceived to move more slowly than uniform targets when moving at high speeds and high contrast.	**Speed misjudged when fast and prey are of high contrast.** *In other situations no effect seen.* *No effect of striping.*
von Helversen, Schooler, and Czienskowski (2013)	
Striped targets perceived as moving faster than uniform objects. Longitudinally striped targets hit more.	**Speed misjudged.** *Cost to longitudinal stripes.*
Hughes, Troscianko, and Stevens (2014)	
Striped targets more difficult to catch than some but not all targets. Striped targets attempted earlier.	*Patterned targets difficult, not stripes specifically.* *Gray as easy to capture as striped objects.*
Hughes, Major-Elliott, and Stevens (2015)	
Perpendicular and oblique striped targets as easy to catch as gray; parallel striped targets easier. Striped targets in groups easier to capture than gray targets in groups.	*No effect of striping in reducing capture success.* *Motion dazzle does not extend to groups.*

Note: Bold typeface indicates support for zebra stripes acting as motion dazzle; italics denotes an undermining of support for motion dazzle.

during displacements of retinal images, simulations showed that vertical stripes on the flank produce motion signals similar to a wagon wheel seen in cinematic films that appears stationary or moves backward (veridical motion and movement in the opposite direction). Diagonal stripes on the rump generate signals oriented perpendicular to the stripe angle similar to a barber's pole. It is not clear whether these would operate when a predator was viewing moving stripes from a short distance.

I thought it would be worth seeing whether I could inject more realism into these experiments by checking whether I misjudged flight speeds of live animals. I walked toward groups or solitary zebra, impala, topi, and waterbuck taking the following measurements. Using a range finder, I recorded my distance from the nearest individual as I stepped from the car. Then, I deliberately walked toward the group at normal speed, counting my paces, but stopped as soon as the herd bolted. Subtracting the number of paces from the initial distance gave me their flight initiation distance. Now I estimated the angle (to the nearest 10°) at which the group fled from me, and I started a stopwatch. When the group stopped running, I stopped the watch and took a second distance reading. Using the flight distance, estimated angle of departure, and stopping distance from me, I could roughly calculate flight length. Dividing this calculated flight length by flight time generated "calculated flight speed."

During the flight, I made a subjective note of whether I thought the flight was very fast, fast, medium, slow, or very slow (which I termed "subjective categorizations"). I also recorded estimated flight length as I judged at the time, and dividing this by recorded flight time, I could calculate a measure called "subjective flight speed." I reasoned that if zebras appear to run slower than nonstriped species as predicted by most versions of motion dazzle, my two personal estimates of flight speed should be lower than those of other species.

"Calculated flight speeds" revealed that zebras ran on average at 3.94 m/ sec (N = 60 flights), whereas waterbuck fled at 4.70 m/sec (N = 21), impala fled at 4.98 m/sec (N = 95), and topi fled at 5.72 m/sec (N = 18); zebras fled from me at significantly slower speeds than impala and topi [12]. I found that the proportion of flights that I categorized as very slow, slow, medium, fast, and very fast (see Figure 4.1) also differed by species, with zebra being subjectively classified as running slower than the other three species [13]. So my impressions generally conformed to species' differences in "calculated flight speeds." When I directly compared "calculated flight speeds" with my four-point subjective categorization (lumping fast and very fast), there was no significant difference, indicating that indeed my subjective categorizations were reasonably good. There was a significant difference across species, however, and a categorization × species significant interaction [14], which was driven entirely by zebra actually fleeing more slowly than impala and topi.

Turning to the second subjective measure of "subjective flight speeds," I found significant differences across species driven by my judging topi to run faster than zebra rather than zebra running slower than the rest [15]. When I subtracted "subjective flight speed" from "calculated flight speed," I found that I was actually judging speeds rather more realistically for zebras than for other species (figures closer to 0, Table 4.4), although these differences did not differ significantly between species [16]. In short,

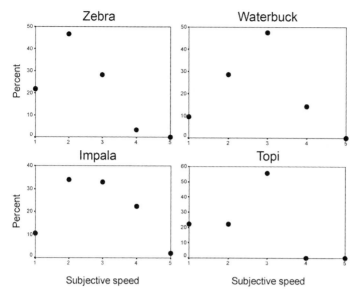

Figure 4.1. Subjective categories of flight speeds from left to right (1, very slow; 2, slow; 3, medium; 4, fast; 5, very fast) of different zebra (N = 60 groups), waterbuck (N = 21), impala (N = 94), and topi (N = 18) groups running away from me.

Table 4.4. Flight speeds of species fleeing from the author in Katavi National Park.

	N	X	SD
Estimated flight speeds			
Zebra	60	3.94	1.71
Impala	95	4.98	2.28
Topi	18	5.72	3.12
Waterbuck	21	4.70	2.37
Subjective flight speeds			
Zebra	51	3.19	1.03
Impala	94	3.71	1.63
Topi	18	4.25	1.83
Waterbuck	19	3.64	1.04
*Difference between estimated and subjective flight speeds**			
Zebra	58	0.75	1.47
Impala	91	1.27	2.10
Topi	18	1.47	2.84
Waterbuck	19	1.22	1.70

*Average calculated flight speeds minus subjective flight speeds of different herbivores fleeing from me where positive values indicate that subjective flight speeds underestimated calculated flight speeds.

I judged all species to be running away from me more slowly than they actually were, but I was certainly not doing this disproportionately more for zebras.

While motion dazzle may operate under restricted circumstances (on computer screens using moving targets) and at one time was widely acknowledged to cause the speed and angle of boat trajectories to be misjudged at sea, very preliminary evidence presented here suggests that it does not cause a human observer to misjudge the speed of a striped herbivore more than nonstriped herbivores under field conditions. My observations do not support a motion dazzle effect in fleeing zebras.

4.7 Misjudging the size of prey

4.7.a Subjective and estimated heights and girths

The Helmholtz illusion (Helmholtz [1867] 1962) refers to a square composed of horizontal lines that appears taller and narrower than an identically sized square made up of vertical lines. P. Thompson and Mikellidou (2011) showed that a rectangle of vertical stripes needs to be extended by 7.1% vertically to match the perceived height of a similar-size square of horizontal stripes and that a rectangle of horizontal stripes must be made 4.5% wider than a square of vertical stripes to match its perceived width. The Helmholtz illusion operates for women's dresses whether the horizontal or vertical lines are on the dress of a line drawing of a woman or on a three-dimensional mannequin viewed stereoscopically. Naively, then, zebras might appear shorter when viewing their flank but taller when viewing their rump as compared to a uniformly colored animal.

Cott (1940) discussed a related illusion in which contrasting elements, not necessarily angled in vertical or horizontal directions, but that interrupt or are close to the margin of an object, make it look larger than those objects where the pattern conforms to the border. He expounded on this using examples of a square, a rectangle, and a circle but also illustrated it with a zebra, implying that it would appear larger than it really was (Figure 4.2). Therefore, it is possible that the vertical stripes on the flank and neck of a zebra might make it appear larger (following Cott) or shorter and fatter (following Helmholtz) than a nonstriped animal. Horizontal stripes might make a zebra appear taller and thinner (following Helmholtz) or have a fuller figure (following Cott) than a uniformly colored animal.

To investigate these possibilities with live zebras, I scored individual

Figure 4.2. Cott's (1940, Figure 40) drawing illustrating how stripes perpendicular to the body's outline (right) may make an animal look larger. He used the same drawings to illustrate how the markings of a zebra could potentially disrupt its outline compared to an animal with internal markings (Chapter 2).

adult and subadult zebras and nonstriped topi and waterbuck in the field. I used a 5-point scale of subjective heights at the shoulder ranging from short to tall, and a 5-point scale of girths ranging from thin to fat; pregnant females were scored as 5. If zebra stripes tricked me into believing that individuals were taller or fatter than they really were, my scores should have erred toward the high end of the 5-point scale, and this seemed to be the case (Figure 4.3), because I scored more individuals in the medium-tall and medium-fat categories than would be expected by chance [17]. Yet although I was overly generous in my subjective size estimates for zebras, I was no more generous for them than for topi and waterbuck (although numbers of waterbuck were very few indeed; Figure 4.3). There were significant species differences between my subjective shoulder height estimates, but this is because I scored most topi as tall, not because I scored striped zebra as taller than other species [18]. There were no differences in my girth estimates across species [18].

In order to compare my subjective heights and girths to the real heights and girths of each individual, I additionally took a photograph of the left-hand side of the body with a Nikon D50 digital camera mounted with a 70-300 AF Nikkor zoom lens 1-4-5.6G always set at 300 mm and immediately noted the distance from the camera with a range finder. Photographs were subsequently imported into Adobe Photoshop CS3. (When I took repeated photographs of the same individual over time, I used the best photograph or, if equally good, the first photograph for subsequent examination.) I

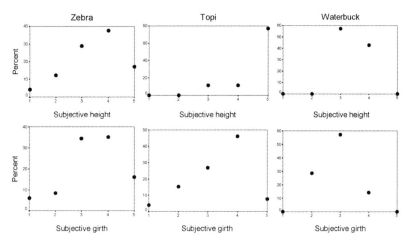

Figure 4.3. Percentage of individual adult and subadult zebra (N = 287), topi (N = 26), and waterbuck (N = 7) that I scored subjectively on a 5-point scale for shoulder heights (1, short; 2, medium short; 3, medium; 4, medium tall; 5, tall) and girths (1, thin; 2, medium thin; 3, medium; 4, medium fat; 5, fat).

pinpointed the following areas of the body using the x- and y-axis tool in Photoshop: top of the scapula, point on the chest distal to where the foreleg emerges from the body, front hoof, hip, groin where the hindleg disappears into the trunk, a point immediately behind the ear, and the "midflank." The last point was obtained by placing a ruler between the scapula and chest points and recording the midpoint between the two. These points allowed me to calculate shoulder height and weights of each animal objectively.

Ostensibly, shoulder height would be the distance between the scapula to the ground, but since the front hoof nearest to me might not necessarily be placed vertically below the shoulder, I used this formula to calculate shoulder height (SH):

$$SH = \sqrt{([\text{scapula x} - \text{hoof x}]^2 + [\text{scapula y} - \text{hoof y}]^2)}.$$

I then multiplied this figure by the distance from the camera and divided by 300 (following Shrader, Ferreira, and van Aarde 2006).

A horse's estimated weight is its girth (G), actually measured as half girth in the field because of tape slippage and then multiplied by 2, which is then squared, and next multiplied by trunk length (L), and finally divided by 228.1 (Milner and Hewitt 1969). I could estimate G at any given distance as follows:

$$G = \left(\left\{ \sqrt{([\text{scapula x} - \text{chest x}]^2 + [\text{scapula y-chest y}]^2)} \times 2 \right\} \times \text{distance} \right) / 300.$$

L was estimated as

$$([\text{scapula x} - \text{hip x}] \times \text{distance}) / 300.$$

I then calculated volume as

$$(G^2 \times L) / 228.1.$$

I found that my estimates of shoulder heights on a 5-point scale were highly correlated with calculated shoulder heights for individuals of each species [19]. My estimated girths were good for zebra but poor for topi and waterbuck, perhaps because I had more practice with zebra or because of the larger sample [20]. Nonetheless, these preliminary findings suggest that I was not disproportionately inaccurate at estimating the size of striped herbivores and fail to support the sixth form of confusion, in which stripes affect judgments of size.

4.7.b Subjective heights and girths and degree of striping

If striping at the margin of the zebra prevents a predator from judging prey's size correctly, one might expect individuals with differing numbers of stripes (perhaps more), or with stripes of differing thicknesses (perhaps thinner), or with differing degrees of white and black pelage to be differentially difficult to judge. I therefore scored stripe characteristics from photographs. I recorded the number of black stripes between the distal point of the ear to the "midflank" whether the neck was above horizontal as when the animal was vigilant or resting, horizontal as when the animal was walking, or below the horizontal when feeding or drinking; in the great majority of cases, it was above horizontal. I also recorded the number of black stripes between the "midflank" point and groin. In both cases, I placed a ruler on the computer screen and, for each black stripe, determined its width using the x-axis tool where the ruler crossed that stripe. Totaling these widths gave me a crude estimate of the extent of black on different parts of the body. Dividing this figure by the number of stripes gave me an average vertical black stripe width for the body. By calculating the lengths of the two vectors between the point behind the ear and "midflank," and "midflank" and groin, I could determine the percentage of those vectors that were black for both neck and flank sections of the animal.

For the rump, I used a slightly different method because rump stripes were so wide and few in number on plains zebras in Katavi. Here I laid the ruler between hip and groin points and recorded the pixel numbers using the y-axis tool where the vector crossed margins of pelage stripes. After calculating the length of this hip-groin vector, I determined the proportion of the vector that was black in the direction of the y-axis.

Finally, I measured the width of the widest black stripe and width of the widest white stripe on the zebra's body, which were usually on the haunch, so I used the y-axis tool. I then calculated the ratio of these widest stripes to each other—that is, (black/[white + black]) × 100. Most measures of striping were intercorrelated. Average width of lateral stripes, ratio of widest black and white stripes, and percentages of neck, flank, and rump that were black were generally highly and significantly positively correlated with one another. The total number of lateral stripes was significantly negatively correlated with average stripe width and positively with increasing ratios of black to white on the rump and with percentage of black pelage on the neck (Table 4.5).

There was no association between either the five subjective shoulder height categories or the five subjective girth classifications and the five measures of striping, save for one marginally negative association between number of (vertical) lateral stripes and my subjective measure of shoulder height [21, 22]. In summary, carefully controlled, properly replicated psychophysical experiments in other contexts show that stripes generate human misjudgments of size, but in the field an almost complete lack of association between aspects of striping and subjective measures of size suggests that my familiarity with zebras, and by extension that of their predators, makes these illusions irrelevant under natural conditions.

Table 4.5. Pearson correlation coefficients and p-values (parentheses) of measures of striping from 287 different plains zebras.

	Average width	B/W ratio	% black on neck	% black on flank	% black on rump
Total lateral stripes	−0.212	0.122	0.243	0.038	−0.052
	(<0.0001)	(0.039)	(<0.0001)	NS	NS
Average stripe width		0.240	0.356	0.250	0.263
		(<0.0001)	(<0.0001)	(0.001)	(<0.0001)
Black/white ratio			0.274	0.353	0.383
			(<0.0001)	(<0.0001)	(<0.0001)
Percentage black on neck				0.149	0.240
				(0.012)	(<0.0001)
Percentage black on flank					0.181
					(0.002)

4.8 Quality advertisement

On the basis of a single experiment with human subjects, it has been suggested that striping patterns advertise some aspect of quality of an individual zebra, either in relation to its ability to win intraspecific contests, in mate choice, or regarding its ability to escape predators (Ljetoff et al. 2007). Specifically, people were asked to pick out a behaviorally deviant object from a group of five similarly patterned objects moving across a computer screen in a sawtooth fashion. Behavioral deviance meant performing some forms of positional and synchronal oddity in the group. The objects were either black, or spotted, or had a black border that graded into a white center, or had horizontal and vertical black and white stripes. Subjects found it easier to identify behaviorally deviant striped objects placed in the middle of the group and could identify these quickest. The authors reasoned that stripes make behaviorally odd individuals more recognizable.

If, in nature, these odd individuals were of low phenotypic quality and thus easy to pursue, it is not clear why they would be selected to advertise themselves with stripes. If they were high-quality individuals, it is difficult to see how quality advertisement could be maintained because individuals of lower quality would also gain, say, antipredator advantages from being striped, and the signal would become dishonest over evolutionary time. Even if these logical obstacles could be surmounted, one would need to show that predators take, or at least hunt, disproportionately more low-quality zebras than low-quality individuals from uniformly colored prey species. These problems notwithstanding, there is considerable intraspecific variation in aspects of striping, and individuals can be recognized by their striping patterns (Peterson 1972; Penzhorn 1984), so it is not impossible that predators could identify and selectively track targets based in part on individual striping patterns.

Neither I nor others have data on whether individuals with certain stripe patterns are targeted more by lions or spotted hyenas. It would take considerable fieldwork time, and many logistical hurdles would have to be overcome to ascertain this, but it seems unlikely a priori that a lion, a stalking predator that needs to get within 30 m of prey to have a chance of making contact, would choose among individuals based on their striping; rather, it would take the closest or least vigilant individual (FitzGibbon 1989). If striping was a form of pursuit deterrence, more likely it would be directed at spotted hyenas that lope toward a herd of zebra and watch them gallop off and then single out the least physically able individual (Schaller 1972), or wild dogs that pursue zebras for 0.3–1.5 km and then pick off the individual that lags behind (Schaller 1972). While anecdotally, anyway, both predator species single out a vulnerable individual based on its flight characteristics rather than on its external coloration, neither carnivore imposes strong predation pressure on zebras (see section 4.10).

4.9 Conclusion

At least six mechanisms have been proposed by which zebra stripes might confuse predators and so allow zebras to reduce attack-kill ratios, yet there is little supporting evidence for black and white striping confusing predators by any means. Miscounting zebras seems implausible, especially since striping would reduce apparent group size, making herds more attractive to predators. Individuals usually flee with gaps of light between them, suggesting striping does not obscure the outline of a fleeing individual. Groups of zebras do not show erratic behavior in flight. My estimates of flight speed and sizes of zebras were not adversely affected by stripes, providing preliminary data that striping does not cause misjudgments of speed or target size under field conditions. Moreover, lions seem to target prey accurately.

Lions are more successful when hunting solitary prey individuals and when hunting very large groups (Schaller 1972; van Orsdol 1984). The latter effect may be related to confusion with individual prey animals moving in many directions but is not necessarily related to striping per se. For example, I have seen what appeared to me to be confusion in cheetahs hunting large herds of wildebeest and in adolescent cheetahs hunting large groups of Thomson's gazelles (Caro 1994). Detailed studies of lions do not report confusion when attacking zebra, however. Indeed, Schaller (1972, p. 255) wrote, "At night, the time when lions do most of their hunting, the stripe pattern loses its contrast. Besides, lions seldom leap at their prey." Some calculations are helpful here. Relative speeds of zebras and lions are known (Jeschke and Tollrian 2007). Maximum escape speeds of plains zebra are 19 m/sec and of lions are 16 m/sec (Garland 1983), giving a relative escape speed of 1.2. Maximum acceleration speeds are 5.0 m/sec^2 for zebra and 9.5 m/sec^2 for lions (Elliott, McTaggart Cowan, and Holling 1977), giving a ratio of 0.5. Lions must therefore catch zebra by surprise within the first 6 seconds of revealing themselves. They probably choose their victim before or during the stalk before breaking cover. Logically, confusion seems unlikely.

4.10 Difficulties with the predation hypothesis

More generally, the hypothesis that striping is an antipredator defense, whether operating through crypsis (Chapter 2), aposematism (Chapter 3), or confusion (Chapter 4), is greatly weakened by the degree to which plains zebra populations suffer from predation, at least in certain ecosystems. For example, predators account for around 30% of the annual mortality of plains zebras in the Serengeti, and the zebra population there may be limited by foal mortality, two-thirds of which may be due to predation (Grange et al. 2004). Examining predator species separately, lions prefer to take plains zebras more than would be expected from their abundance (Plate 4.2). Take, for example, Pienaar's (1969) monumental 20-year study in Kruger National Park, where he calculated that lions take a far greater proportion of zebra than would be expected from their relative abundance (Table 4.6). More convincing still, in a meta-analysis of 40 lion field studies throughout the lion's geographic range, Hayward and Kerley (2005) compared the proportion of total kills made by lions (r) and the proportional availability of prey species (p) using Jacobs's (1974) index, where

$$D = (r - p)/(r + p - 2rp).$$

Resulting D values range from +1 (maximum preference) to –1 (maximum avoidance) and can be tested for significant "preference" or "avoidance" using t-tests or sign tests. Lions (males 150–240 kg, females 122–182 kg) significantly prefer plains zebras (Jacobs's index = +0.16) along with gemsbok, buffalo, wildebeest, and giraffe. The mean mass of significantly preferred prey weight is 290 kg, and most preferred prey weight is 350 kg (see also Clements et al. 2015). Plains zebra weigh 175–385 kg and were recorded as a lion prey item in more studies than any other prey species! Lions encounter zebras more than would be expected by their abundance

Table 4.6. Prey preference ratings (percentage of kills/percentage abundance) of four species of carnivores in Kruger National Park feeding on the five most common prey species.

	Impala	Buffalo	Zebra	Wildebeest	Kudu
Percentage relative abundance	53.4	8.7	8.1	7.8	2.9
Lion	0.37	1.06	1.98	3.06	3.82
Leopard	1.45	0.01	0.15	0.17	1.00
Cheetah	1.27	0.01	0.23	0.65	2.35
Wild dog	1.63	—	0.02	0.05	1.50

Source: From Pienaar (1969).

Note: Scores greater than 1 denote prey taken in greater proportion than its relative abundance.

because lions forage in areas used by zebras, but they hunt them as frequently as expected when they do encounter zebras (Hayward et al. 2011). Lionesses tend to hunt in groups when after zebras (Scheel and Packer 1991), with hunting success reaching 27.0% (Schaller 1972). To conclude, lions' "preference" for zebras across a wide swath of the African continent makes it very difficult to argue that stripes are any sort of effective antipredator defense mechanism against lions. At a stretch one could argue that preferences would be even stronger for zebras in the absence of striping, but the high lion predation rates on similar-size (unstriped) prey as zebras suggest striping would be of little benefit.

Again using Jacobs's index, now comparing 16 spotted hyena (45–80 kg) studies across their geographic range, Hayward (2006) found that plains zebras are statistically significantly "avoided" by spotted hyenas (Jacobs's index of –0.44), along with buffalo and giraffe, while no prey species are significantly "preferred." Spotted hyenas usually take prey with a body mass range of 56–182 kg with a mode of 102 kg, below that of an adult plains zebra. Certainly, some populations of spotted hyenas kill zebras regularly (e.g., Trinkel 2010), but Hayward's analyses are of a more general nature that refer to continent-wide trends. Given that other significantly "avoided" prey, giraffe and buffalo, are far larger than zebras and may be difficult to tackle, but that eland and gemsbok are of greater or of similar body weight to zebras and are not significantly "avoided" by hyenas, it is conceivable that stripes operating through crypsis, aposematism, or confusion result in zebras being "avoided" by this species of carnivore, although it is more parsimonious to attribute this to their relative body size, because they are outside the range normally taken (Owen-Smith and Mills 2008) or preferred (Clements et al. 2015) by this predator.

Adult zebras are also captured by wild dogs and very occasionally by leopards and cheetahs, but zebras are very large prey items for these smaller carnivores. Wild dogs (17–36 kg) "prefer" 16–32 kg prey and also 120–140 kg prey. Their Jacobs's index for plains zebra is –0.88, statistically significantly less than expected based on their abundance, and it is –1.00 for mountain zebra (Hayward, O'Brien, et al. 2006). For leopards (20–90 kg), their "preferred" prey weight is 10–40 kg, and they have a Jacobs's index of –0.80 for plains zebra, significantly less than expected, and –1.00 for mountain zebra (Hayward, Henschel, et al. 2006). For cheetahs (30–72 kg), their "preferred" prey weight is 23–56 kg, far below that of subadult or adult zebras. Cheetahs' Jacobs's index for plains zebra is –0.69, which is significantly "avoided," and –1.00 for mountain zebra (Hayward, Hofmeyr, et al. 2006). Adult zebras are generally too large to be tackled by these carnivore species, although there may be exceptions—for example, one pack of wild dogs in the Serengeti was a zebra specialist (Malcolm and van Lawick 1975). The most parsimonious explanation for midsize carnivores avoiding zebras is their relatively small body sizes rather than stripes being an antipredator device (Owen-Smith and Mills 2008; Clements et al. 2015). If stripes were an antipredator mechanism—and most of the data in these three chapters point against it—surely it could only be in relation to spotted hyenas, the most numerically abundant carnivore and the one on the cusp of being able to handle prey of zebras' body weight. Any remaining diehards adhering to an antipredation function of stripes would do well to study the hunting behavior of spotted hyenas.

Statistical tests

[1] Univariate model comparing standard errors (variability) of counts of three observers during morning and evening watches combined separating the data according to 4 group sizes 1–10, 11–50, 51–100, and 101–200 animals for each of the 10 species: group size $F = 109.797$, df = 1, 179, $p < 0.0001$; observer $F = 3.142$, df = 1, 179, $p = 0.077$; species $F = 0.266$, df = 9, 179, NS.

[2] Multiple logistic regression on whether groups bunched together during flight from me against flight initiation distance (Wald = 0.246, df = 1, NS), number fleeing (Wald = 0.066, df = 1, NS); flight time (Wald = 0.543, df = 1, NS), and flight speed (Wald = 0.018, df = 1, NS).

[3] Binary logistic regression on whether I could see a more or less continuous line of stripes against flight initiation distance (Wald = 0.991, df = 1, NS), number fleeing (Wald = 0.071, df = 1, NS); flight time (Wald = 0.072, df = 1, NS), and flight rate (Wald = 4.015, df = 2, NS) from me.

[4] Multiple logistic regression on whether the line of stripes was continuous against flight initiation distance (Wald = 1.526, df = 1, NS), number fleeing (Wald = 3.549, df = 1, $p = 0.060$); flight time (Wald = 1.655, df = 1, NS), and flight speed (Wald = 3.017, df = 1, $p = 0.082$) from me.

[5] Multiple logistic regression on no gaps being seen between neighbors at 5 seconds into the flight against flight initiation distance (Wald = 0.012, df = 1, NS), number fleeing (Wald = 0.183, df = 1, NS); flight time (Wald = 0.973, df = 1, NS), and flight speed (Wald = 4.742, df = 1, $p = 0.029$) from me.

[6] Mann-Whitney U tests comparing the proportion of times that stripes on adjacent zebras with overlapping bodies were not aligned in frames in side view flights with lions absent (16 videos) and with lions present (14 videos), U = 45.5, $p = 0.005$; hind view flights with lions absent (4 videos) and with lions present (6 videos), U = 6.0, $p = 0.068$.

[7] Side views: Mann-Whitney U tests comparing the proportion of times that stripes on adjacent zebras with overlapping bodies had stripes aligned in flights with lions absent (16 videos) and with lions present (14 videos), vertical neck and flank stripes, U = 80.5, NS; horizontal rump stripes, U = 100.5, NS; oblique rump stripes, U = 50.0, $p = 0.010$.

[8] Hind views: Mann-Whitney U test comparing the proportion of times that stripes on adjacent zebras with overlapping bodies had horizontal stripes aligned in flights with lions absent (4 videos) and with lions present (6 videos), U = 6.0, p = 0.068.

[9] Multiple logistic regression on likelihood of fleeing at a sharp angle to me (46–90° from my direction of travel) against flight initiation distance (Wald = 5.925, df = 1, p = 0.015), number fleeing (Wald = 0.212, df = 1, NS); flight time (Wald = 0.454, df = 1, NS), and flight speed (Wald = 0.435, df = 1, NS).

[10] Binary logistic regression on likelihood of zigzagging against flight distance (Wald = 3.895, df = 1, p = 0.048), number fleeing (Wald = 0.367, df = 1, NS); flight time (Wald = 3.722, df = 1, p = 0.054), and flight rate (Wald = 4.849, df = 2, p = 0.089).

[11] Binary logistic regression on likelihood of rear animal switching position during flight against flight initiation distance (Wald = 0.012, df = 1, NS), number fleeing (Wald = 1.928, df = 1, NS); flight time (Wald = 0.661, df = 1, NS), and flight rate (Wald = 0.909, df = 2, NS).

[12] Calculated flight speeds across species F = 4.091, df = 3, 190, p = 0.008. Tukey HSD tests, zebra vs. impala mean difference = −1.037, p = 0.024; zebra vs. topi mean difference = −1.778, p = 0.015; zebra vs. waterbuck mean difference = −0.761, NS.

[13] Number of flights classified by subjective category as very fast and fast combined, medium, slow, or very slow differed by species F = 5.043, df = 3, 189, p = 0.002. Fast and medium vs. slow and very slow, zebra vs. impala χ^2 = 8.244, df = 1, p < 0.01; zebra vs. topi χ^2 = 3.552, df = 1, p < 0.1; zebra vs. waterbuck χ^2 = 6.504, df = 1, p < 0.02.

[14] General linear model of calculated flight speeds across species and subjective categorizations Model F = 2.446, df = 14, 175, p = 0.004, subjective category F = 0.752, df = 3, 175, NS, species F = 4.411, df = 3, 175, p = 0.005, subjective category × species F = 2.132, df = 8, 175, p = 0.035. Tukey HSD tests, categories fast and very fast combined vs. medium, mean difference of calculated flight speeds = 0.169, NS; fast and very fast combined vs. slow, mean difference of calculated flight speeds = 0.449, NS; fast and very fast combined vs. very slow, mean difference of calculated flight speeds = −0.450, NS; zebra vs. impala mean difference of calculated flight speeds = −1.096, p = 0.013; zebra vs. topi mean difference of calculated flight speeds = −1.778, p = 0.013; zebra vs. waterbuck mean difference of calculated flight speeds = −0.761, NS.

[15] Calculated subjective flight speeds across species F = 2.958, df = 3, 185, p = 0.034. Tukey HSD tests, zebra vs. impala mean difference = −0.518, NS; zebra vs. topi mean difference = −1.056, p = 0.033; zebra vs. waterbuck mean difference = −0.449, NS.

[16] Calculated flight speeds minus calculated subjective flight speeds across species F = 1.043, df = 3, 182, NS.

[17] Subjective shoulder heights on a 5-point scale for 287 individual adult and subadult zebras were significantly larger than expected: χ^2 = 101.902, df = 4,

p < 0.0001. The same was true for subjective girths: $\chi^2 = 113.401$, df = 4, p < 0.0001.

[18] Subjective shoulder heights on a 5-point scale comparing zebra, topi, and waterbuck Kruskal-Wallis test: $\chi^2 = 31.020$, df = 2, p < 0.0001. Subjective girths on a 5-point scale comparing zebra, topi, and waterbuck: Kruskal-Wallis test $\chi^2 = 3.560$, df = 2, NS.

[19] Spearman rank order correlation coefficients of subjective shoulder heights on a 5-point scale and calculated shoulder heights using Photoshop, dropping photographs where the front hoof was obscured by grass or was in mud, zebra N = 206, $r_s = 0.291$, p < 0.0001; topi N = 20, $r_s = 0.567$, p = 0.009; waterbuck N = 6, $r_s = 0.828$, p = 0.042.

[20] Spearman rank order correlation coefficients of subjective girths on a 5-point scale and calculated weights using Photoshop, dropping photographs where the trunk was turned slightly from the camera, zebra N = 262, $r_s = 0.187$, p = 0.002; topi N = 22, $r_s = -0.296$, NS; waterbuck N = 7, $r_s = 0.458$, NS.

[21] Spearman rank order correlation coefficients of subjective shoulder heights on a 5-point scale and measures of striping, excluding photographs where the zebra had twisted its head to look at me, had its neck down, or where the trunk was turned slightly from the camera, N = 180 individuals, number of lateral stripes $r_s = -0.132$, p = 0.077; average stripe width $r_s = -0.036$, NS; ratio of black to white $r_s = 0.024$, NS; percentage black on neck $r_s = -0.090$, NS; percentage black on flank $r_s = 0.077$, NS; percentage black on rump $r_s = -0.045$, NS.

[22] Spearman rank order correlation coefficients of subjective girths on a 5-point scale and measures of striping, excluding photographs where the zebra had twisted its head to look at me, had its neck down, or where the trunk was turned slightly from the camera, N = 180 individuals, number of lateral stripes $r_s = -0.101$, NS; average stripe width $r_s = -0.001$, NS; ratio of black to white $r_s = -0.109$, NS; percentage black on neck $r_s = -0.084$, NS; percentage black on flank $r_s = -0.080$, NS; percentage black on rump $r_s = 0.031$, NS.

Figure 5.0. Tabanid flies inflict painful bites on humans and horses. They avoid landing on striped surfaces. (Drawing by Sheila Girling.)

FIVE

Ectoparasites

5.1 Biting flies

The possibility that white and black stripes evolved in response to biting fly annoyance strikes many people as extraordinary because most think that large predators are the chief selection pressure acting on ungulate prey in Africa. Yet observers of equids note that daily and seasonal movements are often associated with biting fly annoyance (Rubenstein 2011). Moreover, a body of experimental evidence first collected 85 years ago indicates that biting flies are not attracted to black and white striped surfaces (R. Harris 1930; Barass 1960). So what are the mechanisms by which biting flies detect their hosts? In the Glossinidae (tsetse flies), Tabanidae (horse- and deerflies), Muscidae (stable flies and horn flies), Simuliidae (blackflies), and Cuculidae (mosquitoes) families, insects are first attracted to their hosts through odor, vision, and movement, with the relative strength of these factors varying by dipteran (fly) family (G. Gibson and Torr 1999). Carbon dioxide is important for locating hosts from as far as 90 m away for all groups (Zollner et al. 2004), as is octenol, various phenols, acetone (Torr 1989), and ammonia in aged horse urine (Baldacchino, Manon, et al. 2014), but vision is thought to take over at about 20 m from the host. Now, flies make use of color contrast, relative brightness of the host compared to the background, object movement, size, and possibly shape. Most biting fly groups prefer uniform objects and colors in the ultraviolet, blue, and green parts of the spectrum, but, critically for this hypothesis, striping additionally affects behavior of some of these groups (Lehane 2005).

Four studies in particular are central to our understanding of stripes in host-seeking behavior. Working in the field in northern Zimbabwe, Waage (1981) used a 37 cm × 53 cm model painted with 5 cm wide white and black stripes set at a 45° angle, over which he attached electric wires, and next to which he placed a 1 m² electric screen. Arriving flies would be stunned either on the model or on the screen. Uniform black or white models acted as controls. Some models were stationary (8 replicates), whereas others were dragged 4 km at a walking pace (9 replicates), and experiments were conducted without and with (7 replicates) carbon dioxide and acetone attractants. Waage found significantly more *Glossina morsitans* and (principally) *G. pallidipes* tsetse flies landing on both black and white control models than on the striped model whether the models were moving or stationary (Table 5.1a), although these effects were overridden when odor attractant was used with stationary models; numbers of tabanids were too few to test. Since the proportion of flies (the vast majority of which were tsetse flies) electrocuted on the moving model itself compared to the screen did not differ by model color, Waage surmised that striping does not interfere with the landing response but with attraction from some distance. He thought that stripes must reduce attractiveness by obliterating the body edge or reduce the amount of contrast of the animal's form against its background some way off, or else be too narrow themselves to elicit attraction closer up. He acknowledged that olfactory attractants are involved but that the relationship between them and visual stimuli was complex.

In the laboratory, Brady and Shereni (1988) presented satiated male *Glossina morsitans* and *Stomoxys calcitrans* with a paired choice of targets on two walls. A drum rotating for 5 min but then remaining stationary for 3 min was used to activate the flies. Many experiments were conducted in which both the widths and angles of black and white stripes were varied. These researchers discovered that *Glossina morsitans* preferred larger black targets due to their longer edge, males preferred to land on vertical stripes than on horizontal stripes, distinguishing between these when the slope was >50° from vertical. Critically, *Glossina morsitans* demonstrated a reduced landing response as target stripe widths became thinner, particularly noticeable in the targets with 4 and 5 stripes (see Table 5.1b). *Stomoxys calcitrans*, however, was not attracted by large black targets, but it was attracted by their edges, preferred to land on a horizontal than on a vertical stripe, and also exhibited a graded decline in landing with stripe width (Table 5.1b). Brady and Shereni speculated that when two edges are sufficiently close together, neither is seen, with about 8 degrees of separation being the critical point at which the object cannot be seen. Because

Table 5.1. Summary of key findings from experiments conducted on striping and biting flies.

(a) Waage (1981): moving targets with oblique stripes ($\bar{X}$ number of flies/replicate).

	Black	White	Striped
Tsetse	47.7	42.9	26.1
Tabanids	0.6	0.6	0.2

(b) Brady and Shereni (1988): $\bar{X}$ number of landings/active male fly/min × 10³ on moving and static vertically striped targets.

Stripe width (cm)	15	7.5	6.0	3.8	3.0
Stripe number	1	2	3	4	5
Glossina	127	169	116	74	51
Stomoxys	180	66	155	138	52

(c) G. Gibson (1992): captures on single static target ($\bar{X}$ numbers of flies/day).

				Striped	
	Black	White	Gray	Vertical	Horizontal
G. pallidipes	422	558	173	108	8
G. morsitans	10.2	7.1	4.2	4.0	0.5

(d) Egri, Blaho, Kriska, et al. (2012): static salad oil–filled tray targets, total over 2 months.

	Painted containers filled with salad oil				
	Black	2 stripes	6 stripes	12 stripes	White
Tabanids	145	138	66	24	3

their subjects were satiated, they thought that their experiments related to resting, not host-seeking behavior.

Working in the field, again in Zimbabwe, G. Gibson (1992) used stationary electrocuting white, black, gray, and black and white vertically and horizontally striped (5 cm wide) 1 m² cloth targets together with carbon dioxide and acetone attractants. Cloth targets were each presented 4 times either alone or in pairs, and a transparent electrocuting target was placed next to each. Table 5.1c shows that gray and striped single targets but particularly horizontal striped targets were less attractive than white or black cloths for both *Glossina pallidipes* and *G. morsitans*, and these species appeared to avoid horizontal stripes because more were found on a nearby transparent target. In choice experiments, gray was as attractive as black or white, and striped targets captured few tsetse flies when paired with black. She argued that detection of edges of a target perpendicular to stripe orientation, such as the legs of a zebra, would not be enhanced by lateral inhibition and so would be more difficult to detect than edges that lay parallel to the stripes; this could explain why horizontal stripes were more effective in preventing landings than vertical stripes. She also argued that stripes specifically affect the landing response because G. pallidipes can resolve the widest stripes of a plains zebra at 3.0 m and the narrowest at 0.3 m because their maximum resolvable spatial period is 3.5° (G. Gibson and Young 1991).

In the latest series of field experiments, this time in Hungary, Egri, Blaho, Kriska, and coworkers (2012) set out plastic trays filled with transparent salad oil to trap tabanids. The bottoms of trays were painted black or white or with different numbers and thicknesses of white stripes on black backgrounds, but the area of black and white was approximately the same in each. They also set out horizontal sticky striped boards and, in another experiment, vertical sticky boards with a checkered pattern or with stripes of different thicknesses and at different orientations. Overall, they found that trays were less attractive the greater the number of white stripes (Table 5.1.d), that fewer tabanids were captured on sticky white than sticky black stripes, and that stripe widths of 0.23–7.47 cm were least attractive to tabanids. Half-size horse models that were brown, black, white, and striped black and white were also set out. These captured a total of 332, 562, 22, and 8 tabanids, respectively. This research group also explored the effects of polarized light in attracting tabanids, which I will discuss at the end of the chapter.

Taken together, these four studies consisting of numerous experiments are persuasive in suggesting biting flies eschew striped surfaces, but they suffer from problems of experimental design and relevance to the question of why zebra coats are striped. Waage's (1981) study is perhaps the most convincing in regard to glossinids because there were adequate replicate presentations, although the targets were painted cylinders, not zebra pelts. Brady and Shereni's (1988) study is problematic, as the flies were satiated, and so their behavior was likely related to resting rather than host seeking. G. Gibson's (1992) study had few stimulus replicates and used cloth targets that may have different reflective properties than pelage. While Egri, Blaho, Kriska, and coworkers (2012) conducted numerous experiments involving salad oil and painted boards and model horses, these materials also differ from zebra pelage; some of their stimuli were placed horizontally, an aspect generally different from that of a zebra approached by an incoming tabanid; and there were few target replicates of each set of stimuli. Whether these quibbles would substantially change the findings that glossinids, tabanids, and perhaps *Stomoxys* are less likely to land on striped surfaces, and the implication that zebras benefit as a consequence, await more ecologically relevant experimentation.

5.2 Behavioral indices of fly infestation in Katavi

These criticisms notwithstanding, the four sets of studies are provoking because they provide an explanation for striping per se, but they cannot mimic the stimuli given off to flies by live animals, whose particular movements, exhalation, and body odors may influence ectoparasite attraction and landing behavior. I wanted to see whether fewer flies landed on live zebras than nonstriped species, but I could not see flies through binoculars even at close range in Katavi. Instead, I reasoned that if black and white stripes are a deterrent to biting flies, then zebras should be less annoyed by flies than sympatric species living in the same habitat. I therefore watched seven species of herbivore on grasslands from a vehicle and recorded the number of times they swished their tails, since this is known to be a crude measure of insect harassment (Keiper and Berger 1982), and groomed themselves. All observations were made during a short window of time between 28 July and 7 September 2010 under clear skies principally on the Katisunga floodplain in Katavi National Park. I took the following measures: outside temperature in the sun, wind speed on a 5-point scale (from 0 completely still to 4 windy), group size, and main activity—namely, feeding, vigilant, moving, feeding + moving, or vigilant + moving. I watched one focal animal from a group for 3 min before moving my attention to another group. I recorded the total number of times that an individual swished its tail to the right side of its body. I also scored the number of times the individual nibbled its pelage, shimmied its skin, groomed its skin, shook its head, stamped either foreleg or hindleg, scratched its head with its hind hoof, and any other grooming movements and added these together to give a measure that I termed "being bothered by flies" (see Ralley, Galloway, and Crow 1993), although I could not actually discern the flies. Other studies show such behaviors are associated with fly annoyance (e.g., Torr 1994; Baylis 1996; Baldacchino, Manon, et al. 2014). I divided these two totals by the number of minutes watched to give a tail swish rate and a being bothered rate.

There were surprisingly few effects of temperature [1], wind conditions [2], group size [3], or behavior [4] on either measure. Exceptions were giraffe and perhaps topi being more bothered by flies at warmer temperatures; buffalo tail swishing more in warmer conditions; waterbuck tail swishing less when it was windier but more in larger groups; zebra being slightly more bothered in larger groups; and topi tail swishing slightly more when feeding + moving, and buffalo slightly less when feeding + moving. Lack of significant effects can be explained in part because I only started watches after 08.30 hours, when the temperature was already beginning to heat up; 85.0% of observations were conducted under conditions of no breeze, slight breeze, or light breeze; 82.7% of observations were made on group sizes ≤10 animals; and in 66.5% of observations the focal individual was stationary, in 27.4% it was partly stationary, and in only 3.4% was it moving throughout. These data contrast with other studies that show a variety of meteorological factors, including temperature, humidity, and wind speed that affect insect activity, including tabanids (e.g., Van Hennekeler et al. 2011; Baldacchino, Puech, et al. 2014). Since my data were nonnormal and could not be normalized easily, and since effects of these variables were few and inconsistent, I dropped these factors and examined rates of tail swishing and rates of being bothered across species using nonparametric statistics.

There were significant differences in rates of tail swishing and being bothered between species (Table 5.2) [5], with zebras showing higher rates of tail swishing than any of the other species, and higher rates of being bothered than warthog or buffalo [6]. These observations indicate either that zebras are more attractive to flies or that zebras are more sensitive to flies landing on them, or both. If the first, then one could argue that their

Table 5.2. Rates of individual tail swishing per minute and being bothered (by flies) per minute ranked across species in Katavi National Park.

Species	N	$\bar{X}$	SE
Tail swishing			
Zebra	41	40.52	2.54
Waterbuck	18	30.63	5.59
Warthog	17	18.67	3.42
Buffalo	17	10.00	2.20
Topi	22	9.55	1.54
Impala	38	8.12	1.52
Giraffe	26	4.33	0.75
Being bothered			
Zebra	41	1.22	0.26
Waterbuck	18	1.19	0.28
Giraffe	26	1.17	0.53
Topi	22	0.91	0.34
Impala	38	0.67	0.16
Buffalo	17	0.29	0.13
Warthog	17	0.20	0.15

coat coloration does not protect them from flies in general, although their coats could just be aversive to particular families of ectoparasites. If the second, then one could argue that zebras have to resort to both behavioral and morphological strategies to reduce flies landing on them. Certainly, these data indicate that zebras are not free of fly annoyance in parts of their range (see also Joubert 1972) and that stripes cannot be entirely effective in deterring insects in general.

5.3 Behavioral indices of fly infestation in Berlin

In the field I was comparing perissodactyls with artiodactyls, and it is possible that equids tail-swish more and show greater annoyance because they are particularly susceptible to diseases carried by biting flies, or suffer greater fitness consequences of blood loss than do artiodactyls, not because more flies are landing on them. I thought that if I could simply compare members of the equid family, species of which are more likely to be similarly susceptible to parasites or blood loss, I might be able to determine behaviorally whether stripes per se deter flies. I therefore compared the extent to which striped zebra species and nonstriped equids were bothered by insects in the Tierpark Zoo in Berlin, which to my amazement houses all seven species of living equids. Again, I took two measures: rates of tail swishing and rates of attempts to dislodge ectoparasites.

Between 3 June and 2 July 2012, I watched every adult, subadult, and juvenile equid in the zoo for 14 days for a minimum of 3 min each day. I learned to recognize all individuals of each of the seven species by their striping patterns (in the zebras), by individual marks (Przewalski's horses), by patterns of pelage molt (kiang and kulan), or leg striping in the African wild ass (Figure 1.1). All species were housed separately. A total of 48 adults and juveniles were watched, but infants were excluded because they spent a good portion of the day lying down. I recorded the number of times that an individual swished its tail to the right side of its body using a manual counter. I also recorded the number of times the individual was bothered by flies (a total of the number of times the individual nibbled its pelage, shimmied its skin, groomed its forelegs or rubbed its head on its forelegs, shook its head, stamped either foreleg or hindleg, scratched its head with its hind hoof, and any other grooming movements). Instances of mutual grooming were excluded, as they appeared social in nature. I also noted temperature and humidity at the start and end of each watch and for each measure averaged these two readings, and I noted wind conditions on a subjective 10-point scale since these might affect flying insect activity.

The species' enclosures were close together, but to check whether any differences in tail swishing were due to local differences in fly abundance in each enclosure, four fly traps were placed near each enclosure (but out of sight of visitors). These were an odor trap used for catching houseflies (Fly Terminator® Pro trap), two yellow sticky traps (EZ Trap® fly trap) for catching smaller flies, and a translucent sticky trap (Bite Free™ stable fly trap) for catching horseflies. These were set up for a total of 11 days toward the end of behavioral observations (24 June to 4 July 2012). In each enclosure, I recorded both the total number of insects captured in all four traps and the number of hymenoptera that could sting.

There was no evidence that zebras suffered less or more attention from insects than brown or tan-colored equid species (Table 5.3) [7]. If flying insects avoided striping patterns, one would expect less tail swishing in the three zebra species than in the brown species, but this was not the case. The species that tail-swished most was the African wild ass (more than 20/min on average), followed by the Grevy's zebra and plains zebra; Przewalski's horse tail-swished least. Plains zebra appeared most bothered by flies, followed by the African wild ass (Table 5.3). Analyses that took temperature, humidity, and wind speed into account revealed significant differences among species on both outcome measures (Table 5.4). Principally, this was driven by very high rates of tail swishing in African

Table 5.3. Mean frequencies of tail swishes and being bothered by insects per individual per minute across seven species of equid over fourteen days in the Tierpark Zoo, Berlin, showing uncorrected rates and corrected rates (uncorrected rates/captures × 100).

			Corrected for insect abundance	
	TS	BB	TS	BB
Striped equids				
Plains zebra (5)	9.8	1.1	2.3	0.3
Mountain zebra (7)	4.7	0.3	1.5	0.1
Grevy's zebra (5)	12.4	0.5	5.7	0.3
Nonstriped equids				
African wild ass (7)	21.4	0.9	13.1	0.6
Asiatic wild ass (6)	7.4	0.7	2.0	0.2
Kiang (10)	6.5	0.7	1.9	0.2
Przewalski's horse (8)	4.0	0.5	1.9	0.3

Notes: Figures in parentheses show number of individuals. TS: rate of tail swishing; BB: rate of being bothered.

Flying insects in the Tierpark Zoo, Berlin, include: Scarabidae, Sarcophagidae, Calliphoridae (*Lucilia*); Calliphorid (yellow setae); Syrphid (large, fuzzy), Syrphid (regular size), Calliphorid (bluebottle), Muscidae, Fanniidae (?), Vespidae, Boreidae (Scorpionfly), Staphylinidae; Blattodea, Lepidoptera (moth), unknown fly. (Notes from Danielle Whisson.)

Table 5.4. General linear models of $(\log_{10} + 1)$ frequency of tail swishing and $(\log_{10} + 1)$ frequency of being bothered by flies comparing the seven species of equids over fourteen days in the Tierpark Zoo, Berlin.

	df	Tail swish		Bothered	
		F	p-value	F	p-value
Corrected model	9	17.769	<0.0001	5.918	<0.0001
Intercept	1	12.858	0.001	1.849	0.177
Temperature	1	34.559	<0.0001	6.586	0.012
Humidity	1	25.916	<0.0001	5.835	0.018
Wind speed	1	3.754	0.056	6.102	0.015
Species	6	9.571	<0.0001	3.428	0.004

wild asses compared to all other species, and by Grevy's zebras compared to mountain zebras and Przewalski's horses [8]. Regarding rates of being bothered by flies, plains zebras were significantly more disturbed than Grevy's and mountain zebras; African wild asses more than Grevy's and mountain zebras; and mountain zebras less than all other species [9]. In short, there was no clear signature of striping for either of these two measures. The equids hardly differ in tail to body length ratios, so this cannot be an explanation for different rates of tail swishing or being bothered (Table 1.1). It is noteworthy that plains zebras tail-swished far more in Katavi than in Berlin, but evidence of being bothered by flies was about the same (compare tables 5.2 and 5.3).

The number of flying insects that were captured in and around the plains zebra, kulan, kiang, and mountain zebra enclosures was greater than the other enclosures. The mountain zebra enclosure had more wasps, hornets, and bees. Tabanids were captured in the Tierpark, two of them around the Przewalski's horse enclosure, and one near the mountain zebra pen; therefore, they were present but in very low numbers, possibly due to cold June conditions.

Different numbers of flying insects in the seven enclosures could not account for lack of behavioral differences in striped and unstriped equids, however. When tail-swishing rates were devalued by the total number of insects at each enclosure, striped species did not differ from brown equids [10]. African wild asses still showed rates of tail swishing more than twice as much as the next species in line, Grevy's zebra, which in turn tail-swished more than twice as much as the next, the plains zebra (Table 5.3). Accounting for flying insect abundance, rates of being bothered were twice as high in African wild asses than in plains zebras, Grevy's zebras, and Przewalski's horses (Table 5.3). In summary, there was no evidence that stripes reduced insect annoyance of equids in Berlin.

Nonetheless, I had the opportunity to look at effects of stripes on insect landings in more detail. I examined whether tail swishing and being bothered by flies were related to aspects of striping in each species of zebra. Using Photoshop, I extracted the same measures of striping as I had in the field, but this time I had photographs of both sides of the body and took an average (see section 4.7.b). There were surprisingly few significant relationships between my two measures of fly annoyance and striping variables using the ratio of the largest black to the largest white stripe; the total number of stripes; the average black stripe width on the lateral surface; and the percentage of black pelage on the neck, flank, and rump. In plains zebra, however, rates of tail swishing were significantly negatively associated with the proportion of black pelage measured along a transect across the rump; thus, individuals with whiter rumps tail-swished more often [11]. In Grevy's zebra, however, rates of tail swishing were significantly positively associated with average black stripe width along the neck and flank combined; thus, tail swishing was more pronounced in individuals with wider black stripes [12]. There was a hint that individuals with more black on the neck were less bothered by flies [12]. In mountain zebra, there was a hint that individuals were bothered more if their widest black stripe was relatively thicker than their widest white stripe [13]. In summary, there was no clear pattern to these results; nonetheless, the numbers of individuals were extremely small.

Since all three species of zebras are striped, I lumped the data together, although I am not very happy in so doing as Grevy's zebras show very different patterns of striping from the other two species (Chapter 1), particularly absence of striping on the ventral surface and complicated whorls on the rump, and some of the plains zebra in the Tierpark Zoo had shadow stripes. Now, I found that rates of tail swishing and being bothered by flies standardized by flying insect abundance were both significantly positively correlated with the ratio of widest black to white stripes [14], and were marginally significantly associated with average black stripe width on the lateral surfaces [14]. So greater annoyance seemed to be associated with a greater extent of black pelage.

The work at the Tierpark Zoo is challenging because it relates to flying insects in Central Europe, not biting flies in Africa, where zebras live. It shows that white and black striped pelage does not reduce the extent to which equids are bothered by flying insects. There was some suggestion that black pelage on zebras makes individuals more attractive to flies, but only 17 individuals were examined, housed in three pens. The extent to which these findings can be extrapolated to thwarting African biting fly annoyance is an open question, but at least these findings indicate that stripes do not deter insects in general and suggest that high rates of tail swishing in plains zebras are related to local conditions in Africa.

5.4 Tsetse fly traps

5.4.a Biconical traps

Returning attention to Africa, where striped equids and tsetse flies are found, I next tried to determine experimentally whether black and white stripes are less attractive to glossinids than uniform colors. In essence, I wanted to replicate the studies of Waage (1981) and G. Gibson (1992). In and around Katavi National Park in areas where I was attacked by tsetse flies, I erected biconical traps to determine the effects of color and pattern in attracting insects. Biconical traps are one of several sorts of traps used to capture tsetse flies (Challier and Laveissiere 1993). Ten circular biconical traps (circumference 168–170 cm, 165–174 cm in height) were constructed out of cloth and wire. The inside of each trap was divided into four quadrats; four circular holes (15 cm diameter) were cut in the top half of each quadrat that allowed flying insects access to the trap. Two traps were constructed of dark blue cloth, known to be particularly attractive to glossinids (Green 1988; Steverding and Troscianko 2004; Santer 2014); two were dark brown throughout; two were light brown outside but black inside; two were light brown throughout; and two consisted of 4 cm wide black and white stripes both inside and out. The brown traps were made to resemble roughly the shades of brown ungulates that live sympatrically with zebras. Traps were hung from the branch of a tree approximately 50 cm from the ground and fastened at the base to prevent erratic swinging in wind (Plate 5.1). I hung and secured sticky flypapers (TAT®) vertically within 0.5 m on either side or on one side of each biconical trap (always the same number for traps in each session) and replaced these with fresh flypapers each time I checked traps.

Since unbaited traps attract glossinids from only 10–20 m away (Dransfield 1984), I added odor attractants (Vale and Hall 1985b; Green 1986) as follows. Next to each trap I placed a block of industrial octenol (in the first year only) (Blue Rhino® Skeeter Vac®) and a 350 ml empty soda bottle filled to one-third capacity with acetone (every session); this had a drilled stopper from which a wick protruded in the first year, but stoppers were removed in subsequent years. At several different sites over the course of 3 years, traps were erected in pairs 100–150 m apart, depending on the position of available trees, with no respect to color save that pairs could

Table 5.5. Dates and habitats where biconical and cloth traps were erected.

Year	Season	Dates	Days	Habitat
Biconical traps				
2006	Wet	25 Nov–10 Dec	15	Thick woodland (W2)[a]
2007	Dry	18 Sept–22 Sept	4	Thick woodland (W2)
2007	Dry	24 Sept–28 Sept	4	Open woodland (WG)
2007	Dry	29 Sept–3 Oct	4	Woodland (W3)
2007	Dry	5 Oct–9 Oct	4	Woodland (W3)
2008	Wet	30 Jan–5 Feb	6	Woodland (W3)
Cloth traps				
2006	Wet	23 Nov–10 Dec	17	Woodland (W3)
2007	Dry	18 Sept–22 Sept	4	Thick woodland (W2)
2007	Dry	24 Sept–28 Sept	4	Open woodland (WG)
2007	Dry	29 Sept–3 Oct	4	Woodland (W3)
2007	Dry	5 Oct–9 Oct	4	Woodland (W3)
2008	Wet	30 Jan–5 Feb	6	Woodland (W3)

[a]Categories from Kikula (1980); see legend for Table 2.1.

not be the same; pairs were placed 300 m apart from other pairs and over the course of 3 years were set up in six different parts of the Katavi ecosystem (Table 5.5). Traps were checked between 1 and 4 times after they were erected, at which time all insects were removed from the traps and flypapers. The numbers of Glossinidae, Tabanidae, and other insect specimens were recorded, but ants were excluded from tallies. Specimens of each morphospecies were collected and subsequently identified to family level, generating an estimate of family richness per trap.

Table 5.6 shows the total numbers of Glossinidae, Tabanidae, and other arthropod morphospecies that I captured in or next to biconical traps (occasional spiders were found on the flypapers). Numbers of glossinids captured were extremely low, although they were certainly present and in large numbers too, judging by those that bothered me. I captured virtually no tabanids in any biconical trap. Across the six sessions, there were significant differences in numbers of Glossinidae, with more being found in blue traps than some of the brown traps [15], as might be expected based on extensive past research (Green and Flint 1986; Green 1988; Kappmeier and Nevill 1999), but there was no effect of stripes. There were no differences of note between traps for tabanids or other arthropods [16]. Appendix 3 shows morphospecies identified to family level that were captured in biconical traps; the blue trap caught the greatest number of families and the light brown trap the least. In short, striped traps did not reduce capture success, but traps were not very effective at capturing insects.

Plate 1.1. The seven extant species of wild equid. (A) plains zebra, (B) mountain zebra, (C) Grevy's zebra, (D) African wild ass, (E) Asiatic wild ass, (F) kiang, (G) Przewalski's horse.

Plate 2.1. Distant plains zebra on the Katisunga floodplain in Katavi National Park.

E

F

Plate 2.2. Life-size equid models used in experiments. Dimensions were: shoulder height 126 cm, chest to rump 188 cm, nose tip to rump 242 cm, height of top of head from ground 173 cm in all cases. (A) plains zebra, (B) typical gray herbivore control, (C) *left:* typical brown herbivore control, *right:* pelts, (D) brown and white striped model, (E) disruptively colored model, (F) Hugh Cott's model (see text).

Plate 2.3. Photograph of author and his son, Barnabas, taking notes and observing models.

Plate 2.4. Line of models and skins used in the 2008 experiments (see Table 2.5).

Plate 2.5. Three pelts erected for nighttime observations. Note the dog's interest on the far right.

Plate 2.6. A small group of plains zebra taken at a real-world equivalent of 16.4 m as they may appear to a human (*a*, *c*, *e*) and lion (*b*, *d*, *f*) under photopic (bright; daylight), mesopic (dim; dusk), and scotopic (dark; moonless night) conditions. Stripe visibility falls off from human to lion vision and as ambient light declines (from Melin et al. 2016).

Plate 2.7. Groups of zebra (*a, b*), zebra and topi together (*c, d*), waterbuck (*e, f*), and impala (*g, h*) on the Katisunga plains in Katavi National Park from a real-world viewing distance of 9 m. Images are scaled such that size variation between species reflects true differences among species. Images are modeled for human (*left panels*) and lion (*right panels*) visual systems under photopic conditions (from Melin et al. 2016).

Plate 3.1. Plains zebra observing a resting lion at the side of the Katisunga floodplain, Katavi National Park.

Plate 3.2. Freshly wounded zebra showing parallel lines of ripped flesh, probably caused by a lion claw.

Plate 3.3. Plains zebra showing a healed wound where the stripes have not realigned perfectly.

Plate 4.1. Zebras at the Katisunga waterhole in Katavi National Park illustrating how observers think stripes might obscure the outline of an individual zebra or make them difficult to count.

Plate 4.2. Lions feeding on the carcass of a plains zebra in Katavi National Park.

Plate 5.1. Biconical tsetse fly trap with black and white striped pattern.

Plate 5.2. Canopy traps. Glossy paint was used to create the (A) striped and (B) black traps.

Plate 5.3. Canopy pelt traps. (*A*) Zebra pelt, (*B*) wildebeest pelt.

Plate 5.4. The author in the horizontally striped suit.

Plate 5.5. The author walking in (*A*) a zebra pelt, (*B*) a wildebeest pelt.

Plate 5.6. Zebra and wildebeest pelts draped over the Land Rover. Osca Ulaya is examining a pelt for glossinids.

Plate 5.7. Region of interest identification. Each zebra was parsed into eight stereotyped regions for analysis, indicated by the magenta rectangles. The angle of each zebra relative was uncontrolled, but most were largely profile views. While every region was not necessarily visible on every zebra, the same region had to be visible for all seven images that comprised a full polarization series.

Plate 6.1. Plains zebra at the Katisunga waterhole in Katavi National Park. Note the individual on the far right grooming the neck of another individual.

Plate 6.2. Plains zebras showing the variation in striping. The individual on the right has thicker black stripes than that on the left.

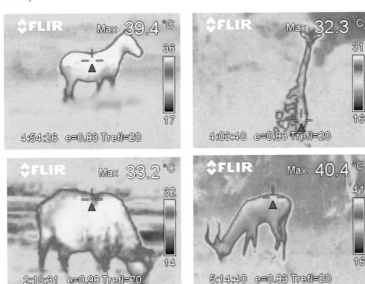

Plate 7.1. Unadjusted thermal images of a plains zebra (*top left*), giraffe (*top right*), buffalo (*bottom left*), and impala (*bottom right*). The triangle marks the location of the maximum temperature in the image.

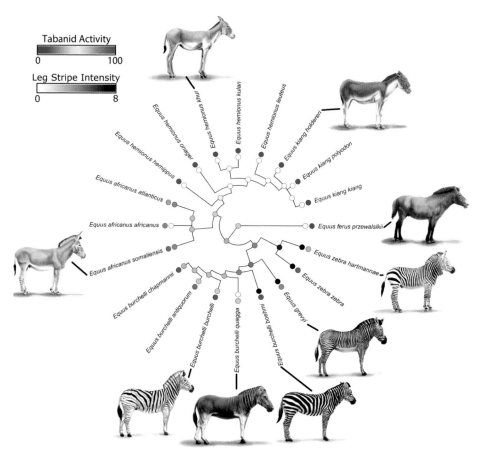

Plate 8.1. Striping and tabanid activity. Phylogenetic tree of equid subspecies showing leg stripe intensity (*inside circles*) and proportion of geographic range overlap with seven consecutive months of temperatures lying between 15°C and 30°C and humidity between 30% and 85% (*outside circles*). (Drawings by Rickesh Patel. From Caro et al. 2014.)

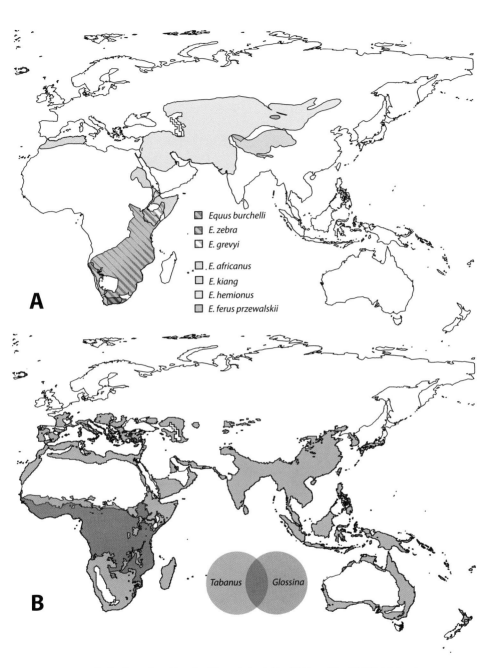

Plate 8.2. Distributions of equids, tabanids, and glossinids. (*A*) Map of the Old World showing distribution of tsetse flies (*Glossina*) and location of seven consecutive months of tabanid activity. (*B*) Geographic distribution of seven species of equids (from Caro et al. 2014).

Plate 8.3. Position of "transects" that were used to measure stripe width on Grevy's zebra pelts in museums.

Plate 9.1. Plains zebras in the dry season in Katavi National Park visiting a waterhole. Note muddied legs, dusty coats, and black splotches on white stripes along flank and dorsa (backs).

Plate 9.2. Two zebra foals traversing the Katisunga floodplain, Katavi National Park. Note long fine brown hairs on the foals' dorsum (back).

Plate 9.3. Two okapi in the Tierpark Zoo, Berlin.

Table 5.6. Biconical traps. Total numbers of Glossinidae, Tabanidae, and arthropod specimens (excluding glossinids and tabanids) that were captured in the wet season in 2006 and 2008 (21 days total) and the dry season in 2007 (16 days total).

Trap color	Glossinidae	Tabanidae	Arthropods
Wet season			
Blue	11	1	498
Dark brown	6	1	569
Light brown / black	6	0	434
Light brown	11	1	355
Black/white stripes	11	0	506
Dry season			
Blue	27	0	3
Dark brown	1	0	4
Light brown / black	4	0	2
Light brown	5	0	44
Black/white stripes	0	0	7

5.4.b Cloth traps

To examine the extent to which tsetse flies landed on white and black striped surfaces compared with uniform black or white surfaces, I conducted a second experiment simultaneously while the biconical traps were erected, because it is common knowledge that trap design affects population estimation considerably (Challier 1982). In this experiment one meter squares of cloth were placed vertically by affixing their corners to the branches of trees and were set 50 cm above ground (Ndegwa and Mihok 1999). I had three jet-black square cloths; three pure white squares; and six black and white striped squares (stripe widths 5 cm), three of which were horizontal and three of which were vertical. I also set up an additional three medium-gray colored cloth traps to resemble sympatric ungulates in the area such as gray waterbuck. These five different colored cloths (N = 15) were erected in a line 300 m apart from one another and were set up in six different areas in the Katavi ecosystem where there were many tsetse flies. Paired sticky flypapers were set up on either side of each square of cloth and a soda bottle filled to one-third with acetone and an octenol block were placed at the base of the tree to which the cloth was attached; the bottle had a stopper and wick in 2006 but not thereafter.

I caught more arthropod individuals using the cloth meter squares than biconical traps (Table 5.7), although numbers of glossinids and tabanids were still very low. Numbers of individuals differed significantly

Table 5.7. Cloth traps. Total numbers of Glossinidae, Tabanidae, and arthropod specimens (excluding glossinids and tabanids) that were captured in the wet season in 2006 and 2008 (23 days total) and the dry season in 2007 (16 days total).

Trap color	Glossinidae	Tabanidae	Arthropods
Wet season			
Black	6	7	1499
Gray	2	4	354
White	2	1	310
Striped (horizontal)	0	0	378
Striped (vertical)	2	1	579
Dry season			
Black	10	10	220
Gray	7	2	123
White	0	2	138
Striped (horizontal)	0	0	137
Striped (vertical)	1	0	128

between traps, with black and gray cloths attracting more glossinids than the others [17] and horizontal stripes catching none. Black attracted more tabanids than white or striped cloths, and interestingly white attracted marginally more tabanids than horizontally striped cloths, which captured zero, but numbers were very low indeed [18]. Black attracted more arthropods in general than other cloths [19].

The numbers of families of insects captured using cloth traps were roughly similar, with most being trapped by the black and horizontally striped traps and least by the gray (Appendix 4).

On reflection, I believe that my attempts to capture glossinids using stationary traps in parts of the Katavi ecosystem where there were many tsetse flies were ill conceived. This is because I subsequently discovered that most of the tsetse flies in and around the park are *G. morsitans*, a species that locates its hosts by movement (Vale 1974). Thus, stationary traps would de facto catch few individuals, as found in G. Gibson's (1992) study. Furthermore, the flypapers next to the cloth traps captured very few large insects, which pulled themselves free if the flypapers were left for more than a few days and the glue thinned in the heat. While I did find that black cloths were more attractive to tsetse flies, as found in many other studies (e.g., Vale et al. 1988), lack of an effect of striping might simply be attributable to small numbers of tsetse flies captured overall. Looking back on this work, collaboration with entomologists or pest control experts would have helped greatly. Next, however, I turned my attention to tabanids.

5.5. Tabanid traps

5.5.a Canopy traps

Tabanids are typically trapped using a hanging object slowly moving in the wind over which is placed a tent up into which they fly (Hribar, Leprince, and Foil 1991). These are called canopy traps. I constructed 18 traps using a small tent made out of white mosquito netting, anchored to 4 posts that were set 1 m apart with a central post to give the center an additional 0.67 m elevation. The skirt of the tent stood approximately 1 m above ground. Hanging from the central pole was a small plastic beach ball approximately 59 cm in circumference. The ball of one pair was painted with glossy black and white stripes (average of mean stripe widths across 6 balls, black $\bar{X}$ = 3.0 cm, SE = 0.2; white $\bar{X}$ = 2.8 cm, SE = 0.2), whereas the other ball was painted entirely glossy black. The balls were covered with Tanglefoot® tree pest barrier, although few tabanids were trapped on balls. Instead, tabanids normally flew up into the net and were captured live by hand. White netting viewed from a distance would tend to dilute differences between balls. An additional two pairs of traps were hanging buckets painted on the outside with black and white stripes ($\bar{X}$ = 2.7 cm, SE = 0; $\bar{X}$ = 2.1 cm, SE = 0.1, respectively) and covered with Tanglefoot®, but they caught only one tabanid. I erected 11 pairs of traps in the dry season of 2013. Members of pairs were separated by approximately 30 m and were placed between 3 and 30 m from standing water and usually on muddy banks (Plate 5.2).

Table 5.8 shows the number of tabanids and glossinids captured over the course of the experiment. There was great variability in capture success across sites, with the three pairs of traps near Lake Katavi (a floodplain with little water in it in the dry season) catching nearly all the flies, but with traps capturing few in other parts of the park. The total number of tabanids (all species combined) caught in the black canopy traps was 107, whereas 13 were caught in the striped canopy traps [20]. The attractiveness of black objects to New World tabanids has been documented in many studies (e.g., Bracken, Hanec, and Thorsteinson 1962; Browne and Bennett 1980; Allan and Stoffolano 1986b), and these findings extend results to African tabanids (here *Tabanus taeniola* and *Atylotus* sp.). In addition, they show forcefully that tabanid abundance varies greatly even within a small area of approximately 50 × 30 km, although my study was conducted only in the dry season, when tropical Tabanidae are less abundant (Lehane

Table 5.8. Canopy traps. Total numbers of captured *Tabanus taeniola*, *Atylotus* spp.—probably *agrestis*, *Glossina* spp. are shown.

Trap		Period of days	Number of checks	*Tabanus taeniola*	*Atylotus*	*Glossina*
1	Black	21	5	8	17	1
	B&W[†]			0	1	0
2	Black	20	5	11	15	5
	B&W			1	2	0
3	Black	21	5	23	35	0
	B&W			2	7	0
4	Black	14	5	0	0	1
	B&W			0	0	0
5	Black	12	3	3	0	0
	B&W			0	0	0
6	Black	15	4	1	1	1
	B&W			0	0	0
7	Black	12	3	0	0	0
	B&W			0	0	0
8	Black	12	3	0	0	0
	B&W			0	0	0
9*	Black	9	3	0	0	0
	B&W			0	0	0
10*	Black	10	2	1	0	1
	B&W			0	0	0
11	Black	6	3	0	0	0
	B&W			0	0	0

Notes: B&W denotes black and white. One *Tabanus par* was caught at Lake Katavi and is included as an *A. agrestis*. (Notes from Steve Mihok.)

Trap locations: #1. side of Lake Katavi, Wameru signboard; #2. side of Lake Katavi, near old ranger campsite; #3. side of Lake Katavi, 1.5 km south of #2; #4. pond north of shortcut to Lake Chada; #5. start of Kavuu River; #6. side of Lake Chada, 0.6 km north of #5; #7. old Ikuu bridge; #8. 0.6 km along North Lake Chada track from new Ikuu bridge; #9. Katuma River, 12.4 km upstream from Sitalike; #10. Palm waterhole; #11. Katuma River, 0.6 km east of old Ikuu bridge.

*No net, two painted buckets covered with Tanglefoot.

[†]Trap fell on one day.

2005). Regarding *Glossina*, numbers were low: eight were caught in the black traps and zero in the striped traps [20]. Importantly, my results show that both families of biting flies eschewed striped balls in comparison to black balls, particularly tabanids, and they do so in Africa, where striped equids are found.

5.5.b Pelt canopy traps

Black and white glossy paint does not necessarily reflect light in the same way as black and white pelage. Certainly this is true of polarized light because, in contrast to paint, mammalian hairs are tubular and lie at different angles depending on where they are found on the body, and so they reflect light at different angles. To make the canopy traps more realistic, I used animal pelts. Specifically, I hung inflated plastic balls of the same dimensions as before under tents made of white mosquito netting also designed in the same fashion. Now, however, the balls were completely covered in either a tight-fitting zebra pelt or a dark brown wildebeest pelt (Plate 5.3). In August 2014, I erected six pairs of canopy traps set 30–86 m apart at 0.8 km intervals along the shores of Lake Katavi, where I knew from the previous year that tabanids were found in abundance. These traps were set out for eight consecutive days. After four days, the pairs of balls were interchanged in order to iron out any differences in trap design, because I noticed in this experiment that more dead tabanids and other flies collected in those traps with a greater degree of folding along each of the four sides of the netting. Flies became entangled in these folds and succumbed to the afternoon heat especially on the west side. To account for trap variation, folds were scored on a two-point scale and summed over all four sides (minimum 5, maximum 8) and were additionally scored on a two-point scale on their western sides; these were entered as covariates in statistical analyses. Additionally, I separated live tabanids found in the apex of the tent from dead tabanids in the folds because the former were less likely to be affected by trap design.

Traps were checked daily between 15.00 and 17.00 hrs during the heat of the day. The number of tabanids captured in each pair of traps differed considerably according to location even over a 4 km distance (Figure 5.1), but, overall, marginally fewer tabanids (the vast majority *Tabanus taeniola*) were captured per day in the zebra skin traps than in the wildebeest skin traps, taking trap construction into account [21]. The reduced attraction of zebra pelts was far more pronounced for the live tabanids than for the dead ones, the latter of which may have been more influenced by differences in trap design (Figure 5.1) [22]. Interestingly, dipteran flies of several species were less likely to be found in the zebra skin traps [23], raising the possibility that zebras are less bothered by nonbiting African diptera than wildebeest. I view this experiment as being the most prescient of all the fly experiments that I conducted in Katavi, since it used standard tabanid traps, placed in an area of known tabanid abundance, and used naturally striped and unstriped mammalian pelage rather than painted objects. This was an experimental breakthrough achieved after a decade of work on zebras in Katavi.

Since the zebra skin surrounding each ball was taken from different parts of the pelt, traps had different numbers of stripes and stripes of different thicknesses. I took photographs of three aspects of the balls and deter-

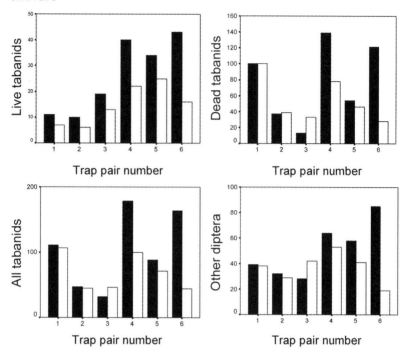

Figure 5.1. Total numbers of tabanids and nontabanid dipterans captured in six pairs of wildebeest (black bars) and zebra (white bars) canopy pelt traps over 8 days.

mined the number of stripes and their widths along a horizontal transect by counting pixels in Photoshop (section 4.7.b). There were no significant associations between numbers of tabanids captured and the proportion of each trap that consisted of black pelage or frequency of black stripes averaged across the three aspects, although the latter correlation was positive and moderately high [24].

5.6 Moving objects

5.6.a Walking in suits

Some biting flies are attracted to movement, especially *Glossina morsitans* (Lehane 2005), and midway through my studies I became concerned that the paucity of tsetse flies caught using biconical and cloth traps might be attributed to lack of motion, an impression reinforced by large numbers of tsetse flies following and alighting on the Land Rover as I moved along the line of traps to check them. Therefore, I had five "suits" of material made for me in Dar es Salaam—black, medium gray, white, horizontally and vertically striped—and used myself as a lure (Plate 5.4). I drove the vehicle to wooded areas inside and outside the national park (in habitat types W2, W3, or WG; see Table 2.1), disembarked, and put on a suit randomly chosen for color. Once the tsetse flies following the car had dispersed, I slowly walked 300 paces in a straight line ahead of and beyond the stationary car dressed in the suit and carrying a sticky flypaper vertically in front of me. I then returned to the vehicle the same way (thereby covering approximately 600 m in total); in the first two sets of walks (i.e., 10), I covered 500 paces in each direction. On my return to the car, Oska Ulaya counted all the tsetse flies and insects that had settled on the suit and additionally any that had alighted on my hat, hands, neck, and face (very few tsetse flies adhered to the flypaper). We then drove 1 km, disembarked again, and I donned a different suit. This was repeated until all five suits had been worn. Walks with each of the five suits were repeated 25 times (a total of 15.8 km in each suit) in the heat of the day.

There were significant differences in numbers of glossinids that were attracted to each suit, with horizontal stripes and black suits receiving most attention (Figure 5.2). Specifically, horizontally striped suits and black suits attracted significantly more tsetse flies than all other suits except each other; and white attracted significantly more than gray [25]. Few of these walks were conducted near water, and only two tabanids landed on me during this experiment: one on the black and one on the white suit. Turning attention to all the other flying insects that were attracted to me, this also differed according to the type of suit that I wore (Figure 5.2). Horizontal stripes attracted significantly more insects than gray, white, or

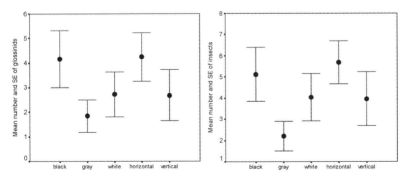

Figure 5.2. Mean (and SE) number of tsetse flies (*left*) and insects overall (*right*) landing on the author when walking 600 m in different colored suits (N = 25 walks in each suit).

vertical stripes; gray attracted significantly less than any other suit [26]. To conclude, moving horizontal stripes on cloth were a cost compared to gray, white, or vertical stripes in attracting tsetse flies.

5.6.b Walking in pelts

Cloth reflects light somewhat differently than does mammalian pelage. For example, saturation on black stripes of the suit was greater than on a zebra pelt, as assessed by photographing both simultaneously at 400 ASA and importing the images into Photoshop ($\bar{X}$ saturation of 10 black stripes = 42.2% and 12.7%, respectively), although not for white stripes ($\bar{X}$ saturation of 10 white stripes = 6.8% and 7.5%, respectively) or for brightness measurements ($\bar{X}$ brightness of 10 black stripes = 24.0% and 26.4%; $\bar{X}$ brightness of 10 white stripes = 73.6% and 83.3%, respectively), which were similar. Cloth also has a different polarization signature from hair (Egri, Blaho, Sandor, et al. 2012). While it was far more convenient to walk in a suit made for me, these considerations forced me to repeat the walking experiments wrapped in a zebra or wildebeest pelt. The protocol was the same, in that I walked 300 m in one direction in woodland (W3 and WG), then returned to the car, where the numbers of glossinids and tabanids were counted on the pelts by Oska Ulaya; in this experiment I dispensed with the flypaper. I repeated these promenades 37 times, walking 22.2 km total in each skin (Plate 5.5).

There was no significant difference in numbers of glossinids that landed on me whether I was wearing the zebra or wildebeest pelt [27]. Tsetse flies were no more likely to land on black or white stripes on the zebra pelt [28]. Marginally more insects landed on me when I was wearing the wildebeest skin [29]. In short, I could find no effect of zebra striping on tsetse flies' willingness to land on a moving primate.

5.6.c Driving with pelts

Some of these walks were conducted inside Katavi National Park within known lion pride territories, and I thought it possible that I might be exhaling more deeply or sweating a little more profusely as I walked over fresh lion tracks 250 m from the car dressed in a zebra skin. I therefore decided to use the car in a choice experiment that also involved movement. Now I strapped a zebra pelt and wildebeest pelt to each side of the car on the bars that normally support the canvas cover of a Land Rover pickup (Plate 5.6). The Land Rover was driven at <10 km/hr for an average of 2.6 km (SE = 0.1, N = 25 drives) through wooded habitats. At the end of each of the 25 drives, I stopped the vehicle, and both Oska Ulaya and I disembarked and then simultaneously counted all the tsetse flies and all insects on each skin at the end, also noting whether they had settled on black or white stripes (Plate 5.6). The advantage of this choice experiment is that the pelts were presented in exactly the same location.

Significantly more tsetse flies and insects were found on the zebra pelt than on the wildebeest pelt [30]; there was no significant difference in the numbers of tsetse flies landing on black or on white stripes [31].

5.7 Conclusions

My experiments and observations in Katavi and elsewhere provide mixed support for striping having effects on biting fly landing responses that range from a strong negative effect with canopy traps, to no effects using cloth traps, to having a positive effect with moving targets (Table 5.9). The experiments suffer from an array of problems, however, which include using cloth material and glossy paint that may not be ecologically valid when testing hypotheses about zebra pelage, and presenting less than ideal numbers of replicates (Table 5.9). The extent to which these are serious problems for interpreting the results is difficult to gauge, but the fact that different groups of insects are involved would make it surprising if all findings went in the same direction. Furthermore, it should be noted that most of the research was conducted in the dry season, when there are relatively fewer arthropods than in the wet season (Table 5.6), which makes it more difficult to uncover differences among stimuli. Moreover, I noted that the distribution of tabanids in Katavi National Park was spatially heterogeneous, with some areas containing high abundances and some none at all. Furthermore, I encountered very few *Stomoxys*, again reducing my ability to uncover differences among stimuli. In addition, *Glossina morsitans*, the predominant tsetse species in Katavi, is attracted to movement, so static traps would not be expected to capture them in large numbers. Last, I was unsure that the concentrations of octenol and acetone were correct (Vale and Hall 1985a) and these affect target attractiveness in glossinids (Bursell 1984; Paynter and Brady 1993). Given these caveats, I now outline what I consider to be the meaningful results.

Considering glossinids first, trap success with stationary traps was low, but I did find that horizontally striped cloths were least attractive to tsetse flies (Table 5.9). Using moving targets, I attracted more tsetse flies to horizontally striped and black suits. These latter results run contrary to G. Gibson's (1992) study, which was also carried out using cloths and movement.

Table 5.9. Summary of my findings broken down by families of biting flies.

Method	Glossinids	Tabanids	Problems
Observations on live animals			
Katavi	Zebras appear bothered by insects		Many confounding factors as compares across species
Tierpark Zoo	No effects of striping on insects		Small sample size and few biting flies
	NA	NA	
Stationary traps			
Biconical traps	No effect but very few	Too few	Few replicates, and material is cloth
Cloth traps	Stripes somewhat less attractive	Horizontal stripes somewhat less attractive	Material is cloth
Canopy traps (balls)	Horizontal stripes less attractive	**Horizontal stripes less attractive**	Glossy paint
Canopy traps (pelts)	None caught	**Zebra pelts less attractive**	None
Moving objects			
Suits	**Horizontal stripes and black most attractive**	Too few	Material is cloth
Pelts on person	**No effect**	Too few	Only two pelts used
Pelts on vehicle	Zebra pelt attracted more	Too few	Only two pelts used; may move between choices

Note: Experiments to which we can attach weight are shown in bold.

One possibility is that she used olfactory attractants—in particular, carbon dioxide—and that this potentiated the attractiveness of unstriped targets. A second possibility is that she principally captured *Glossina pallidipes*, whereas *G. morsitans* in Katavi uses different cues to land on hosts.

I found no differences in numbers of tsetse flies landing on the two sorts of pelts while I was walking. Compared to live animals, pelts were oriented vertically wrapped around me, not horizontally, and this altered stripe and hair orientation. (Additionally, this might have altered the angle of reflected polarized light [Egri, Blaho, Kriska, et al. 2012], although its effects on glossinids are unknown.) There are no previous experiments using pelts to which these results can be compared. More tsetse flies were found on the zebra than on the wildebeest pelt in the driving experiment. Simultaneous presentation of the two stimuli may have confounded results because tsetse flies may have moved between pelts or between car and pelts once they had landed. Furthermore, the loose swarms that characterized tsetse flies following my vehicle may have been related to mating rather than foraging (Brady 1972).

In conclusion, despite my attempts to add realism to the experimental studies carried out by Waage (1981), Brady and Shereni (1998), and G. Gibson (1992) on glossinids, I found it difficult to replicate their results. These differences may stem from differences in trap design (the effects of which are notorious), use of attractants, or because different species of glossinid are found in western Tanzania than in southern Africa.

In contrast to the tsetse experiments, results from the tabanid trapping experiments showed that fewer tabanids were found on horizontally striped canopy traps than on black traps. These findings replicate those of Egri, Blaho, Kriska, and others (2012), who used similar shiny balls to trap tabanids in some of their experiments. Similarly, few were found on horizontally striped cloth traps (Table 5.9). Critically, in my experiment, fewer live tabanids were found on zebra canopy pelt traps than on wildebeest pelt traps. I tentatively conclude from my work that any antiparasite effect of pelage striping in the field may be more related to tabanids rather than glossinids. More generally, I see striping of zebra pelage as being effective in reducing tabanid annoyance and possibly providing the long-sought-after selective pressure driving striping in equids.

5.8 Polarized light

5.8.a Reflected light

Light consists of oscillating electric and magnetic fields lying perpendicular to each other. Each element of light or photon oscillates at a set angle of polarization (E-vector) and will excite a photopigment molecule if the orientation of the excitable double bond in the molecule lies in the same direction of polarization as the photon. Light sources frequently emit photons with varying distributions of polarization angles, and the degree to which light is polarized is called the degree of polarization (d). Sunlight is unpolarized, but as its rays pass through the atmosphere, fine particles scatter blue light and preferentially scatter light polarized in a plane at right angles to the path of rays from the sun. This results in a pattern of polarization determined by the sun's position in the sky, which enables some species to navigate by the sun on cloudy days. Deep, dark waters emit highly and horizontally polarized light, allowing some insects to orient flight toward it (Schwind 1991), whereas shallow and bright alkaline waters reflect both horizontally and vertically polarized light scattered off the bottom or from suspended particles and then refracted at the water's surface (Kriska et al. 2009).

Mammalian hair also reflects polarized light from the surface of the cuticle (shine) and from refraction of incident light within the hair fiber itself and reflection off its back surface, although diffused light scattered by pigments within the cortex of the hair is unpolarized (Lechocinski and Breugnot 2011). According to Horvath and colleagues (2010), white hair reflects light with a lower d than dark brown or black hair because light is backscattered and refracted at white surfaces and is thereby weakly and partially polarized, with direction of polarization perpendicular to the surface, whereas a high d is reflected off black surfaces. Moreover, black hair reflects horizontally polarized light if the light source is low.

Many terrestrial arthropods, crustaceans, and cephalopods can detect polarized light (Horvath 2014), which they may use to navigate, communicate, or detect prey (Horvath and Varju 2004). Within invertebrate eyes, there are long tubes called microvilli covered with a photopigment bearing a membrane that, simply from their geometry, have a 2:1 preponderance of chromophore groups aligned parallel to the long axis; this affords them a capacity to detect polarized light in a given plane. If these microvilli are aligned in different directions, the angle of polarization can be ascertained (Horvath and Varju 2004; Land and Nilsson 2012). In a series of papers, Horvath and colleagues have shown that male and female tabanids are attracted to linearly polarized light. Along with many other insects that lay their eggs in or near water, such as chironomids, mayflies, dragonflies, and some mosquito species (Horvath and Csabai 2014), tabanids are guided by positive polarotaxis to egg-laying sites in or near water. Tabanid eggs are laid on the undersides of leaves or other waterproof surfaces adjacent to muddy or wet sites. After hatching, the larvae drop into moist or dry soil. Adults of both sexes are guided to water as drinking sites and places where they can encounter one another. Female tabanids additionally require a blood meal for egg maturation and use polarotaxis to locate mammalian hosts.

5.8.b Horvath's work

In a long series of experiments, Horvath and his team (2008) set out series of black or white, matte or shiny test surfaces consisting of shiny plastic sheets, sticky plastic sheets, water, or salad oil–filled trays placed horizontally on the ground, or aluminum foil. They found that as many as 27 species of European Tabanidae overwhelmingly preferred to alight on horizontal black shiny surfaces. This was not replicated when surfaces were hung vertically. The shiny horizontal black surfaces reflect horizontally polarized light with a high d. In contrast, shiny horizontal white surfaces reflect unpolarized light, or vertically or obliquely polarized light with a very low d; horizontal matte white, matte black cloth, and brown wooden boards reflect unpolarized light; while vertical black surfaces reflect high d, but this comes off vertically or obliquely. These European tabanids must therefore be attracted to horizontally polarized light. An additional experiment in which sticky horizontal traps were hung 190 cm off the ground facing downward caught no tabanids. This showed that it is the ventral region of the tabanid eye that is polarosensitive (although downward-facing surfaces would reflect less incident light). Interestingly, these ideas had been partially presaged by Bracken, Hanec, and Thorsteinson (1962), who noticed that horizontal cylinders attract more tabanids than vertical cylinders and that smooth, shiny surfaces attract more tabanids than flat or rugose surfaces.

Now they extended these findings to living animals. Observing a brown and a white horse grazing instead of test models, Horvath and coworkers (2010) recorded 3.7 times more tabanids near or sitting on the brown horse. They observed that these dun-colored horses spent more time in the shade, where biting fly attack was reduced. In a series of parallel experiments, they found that tabanids were not attracted to weakly polarizing matte brown cloth or unpolarizing white matte cloth but were attracted to a shiny brown surface with a high degree of horizontal polarization (Horvath et al. 2010). They reasoned that the attraction of brown horses must stem from polarized light, not brightness or color (white or brown). Indeed, imaging polarimetry showed that horizontally polarized light was reflected off the rump, dorsum, and neck regions of brown horses but not off white horses, although this of course depended on direction of view and illumination; other body regions reflected vertically or obliquely polarized light.

In order to explore further the degree and direction of polarized light in attracting tabanids, Egri, Blaho, Sandor, and others (2012) affixed a linearly polarizing sheet to wooden boards and erected these vertically 1 m above ground so that they reflected light vertically, horizontally, or at a 30° angle. Three other boards were placed on the ground with a linear polarizing sheet pointing away, at 90° or at a 45° angle; degree of polarization (d) was the same in all cases. The researchers found no differences in the number of tabanids captured on transparent sticky glue irrespective of polarizing sheet orientation, considering either the elevated or ground-level visual targets that caught more animals. This showed that E-vector direction of polarization was unimportant to the flies. Interestingly, no male tabanids alighted on the elevated surfaces (or on elevated balls), as might be expected for males that search for water but not blood meals on the hoof, whereas both sexes were captured on targets placed on the ground where water is found. The upshot of this experiment is that females must principally use the degree of polarization (*d*) to locate vertical hosts, whereas males and females both look for horizontally polarized light at ground level, as evidenced by earlier experiments.

In a set of experiments key to striping, Horvath's group examined the extent to which striped patterns reduce attractiveness to tabanids. Using trays filled with salad oil, sticky surfaces, or horse models painted with white and black stripes of differing widths, Egri, Blaho, Kriska, and others (2012) found that increasing numbers of stripes trapped fewer tabanids, that black surfaces attracted 2.8–4.7 times more individuals than white surfaces, and that striped horse models attracted fewer tabanids than black, brown, or white models, although numbers were very low in striped versus white comparisons. Striped horse models still caught fewer tabanids than black models when models were baited with ammonia and carbon dioxide (Blaho et al. 2013). Under sunny conditions, black stripes reflected light with a high degree of polarization (*d* > 80% on the horse models), but white stripes did not (*d* < 5%).

In these experiments, polarization was confounded with brightness, but in one further experiment, Egri, Blaho, Kriska, and colleagues (2012) compared boards affixed with 9 linearly polarizing neutral gray parallel strips with alternating orthogonal transmission directions, another with 17, and a third with 17 strips but all with parallel transmission directions. These three surfaces were laid both horizontally and vertically 1 m above ground and covered with sticky glue. The horizontal boards with 17 parallel stripes captured 600 individual tabanids, but those with 17 and 9 alternating stripes caught 208 and 361, respectively. Figures for the vertical boards were 195, 64, and 112, respectively. These results show that fragmentation of the polarization direction alone is sufficient to reduce attractiveness to tabanids irrespective of brightness. Actually, these results are difficult to explain in light of Egri, Blaho, Sandor, and colleagues' (2012) findings showing that degree rather than direction of polarization was important for host finding.

Yet another experiment involved erecting boards vertically with different numbers of black and white stripes set at different angles on the boards, together with a four-square black and white checkered board, and both gray and white control boards (Egri, Blaho, Kriska, et al. 2012). Now, the board orientation meant that horizontally polarized light was only reflected if viewed from one particular direction. Here white vertical surfaces reflecting unpolarized light ($d < 5\%$) captured more flies than gray surfaces ($d < 25\%$) or the checkered surface, demonstrating that brightness must play a role in host choice by tabanids. To summarize all of these experiments, stripe width, polarization direction, brightness (Egri, Blaho, Kriska, et al. 2012), and degree of polarization (Egri, Blaho, Sandor, et al. 2012) are all involved in host-seeking behavior by tabanids, but their relative contributions are as yet unknown and their effects are not entirely consistent across experiments. Moreover, there are issues of ecological validity, with some targets being presented in a horizontal plane rather than vertical as might be experienced by an approaching tabanid, differences between stimulus materials and zebra pelage that make extrapolations risky, and some studies having few stimulus replicates.

Horvath's team also explored the effects of spots rather than stripes on tabanid attraction (Blaho et al. 2012). Similarly to black stripes, brown spots captured 1.4–3.7 times as many tabanids as an equal area of white, and tabanid capture success declined with increasing number and decreasing size of brown spots on both vertical and horizontal test surfaces. Sticky brown cattle models caught more flies than white models with 8, 16, or 64 brown spots; moreover, the 64-spot model captured fewer tabanids than the all-white model. In an experiment with gray linearly polarizing squares affixed to a gray board where squares were either perpendicular or parallel to their surrounding backgrounds, the homogeneous (i.e., parallel) polarization surfaces caught more tabanids than surfaces with four orthogonal spots, which in turn caught more than those with 16 orthogonal spots; horizontal surfaces attracted more than vertical ones. Again, this shows that the angle of polarization signature affects tabanid landing responses independently of brightness. Again, the extent to which intensity of color and polarization are responsible for host attraction is unclear. More generally, this study raises the important question of whether findings from previous studies result from the effects of stripes per se, or whether two-tone patterns are sufficient.

Horvath's extensive studies show that the mechanisms by which tabanids eschew striped surfaces may be related to the pattern of brightness and/or to polarization. The extent to which glossinids avoid black and white striped surfaces using the visible spectrum or polarized light is an open question, however, since although females require a blood meal to reproduce, they do not lay their eggs near water, and it is not known whether parts of their eye are polarosensitive, although this seems probable (Hardie, Vogt, and Rudolph 1989).

5.8.c Polarization signatures of wild zebras

Working with Ken Britten, I wanted to determine directly the extent to which polarized light is reflected off black and white zebra hair under field conditions in Africa. To examine the degree of polarization and plane of polarization of reflected light from both the black and the white stripes on coats of zebras, I took a series of photographs of 21 individual zebras in Katavi. The camera was a Nikon D50 with a 100–300 mm Nikkor lens fitted with a linear/circular polarizing filter designed for digital cameras (Digital Quantaray Professional Filter Series 62mm Circular PL) turned rapidly through seven angles—0°, 30°, 60°, 90°, 120°, 150°, and 180°. From July through to mid-September 2012 in the dry season, images of each subject were captured from a distance of 35–101 m as determined by range finder. Local shading, scene haze, sun azimuth, and elevation were noted for each subject, but in general there was little heat distortion and little overall scene shade. As tabanids forage at approximately 1 m above the ground, and are typically most active in the middle of the day, the time when most photographs were taken, our images would probably be representative of what flies would view when approaching zebras from downwind (Britten, Thatcher, and Caro 2016).

Eight individual regions of interest were localized on the zebra's body: face, foreleg, shoulder, back, dorsum, upper hindleg, hindleg, and rump (Plate 5.7), and we tracked them through the image series. Images were stored in 14-bit/channel raw NEF files (RGB format) and converted by Timothy Thatcher to 48-bit (16 bits/channel) TIFF files using the public-domain software dcraw (v. 9.16, Dave Coffin). The black and white stripes in each region were segmented using a thresholding algorithm that maximized luminance variance between regions relative to within-stripe variance. Polarized light reveals itself as a sinusoidal modulation of luminance with respect to the angle of the filter. The degree of polarization reflects the fraction of emitted light that is concentrated in a particular plane of polarization. A value of 1.0 means that 100% of the reflected light is polarized in a single plane, while a value of 0 results from light containing equal amounts of all polarization angles (as occurs for reflections from matte surfaces).

To determine the magnitude and phase of the polarized light reflected or scattered from each of the eight regions, we fitted the luminance by a sine function of the camera polarization filter setting. The maximum and minimum values of this best-fit function were used to calculate the degree of polarization (d), using the expression

$$d = (L_{max} - L_{min})/(L_{max} + L_{min}),$$

where L_{max} and L_{min} refer to the maximum and minimum luminances, respectively. Michelson contrast between black and white stripes was calculated similarly:

$$contrast = (L_{white} - L_{black})/(L_{white} + L_{black}),$$

where L_{white} and L_{black} denote the average luminance of the white and black regions, respectively (see Britten, Thatcher and Caro 2016 for further details).

We found that both black and white stripes reflect substantial amounts of polarized light and that the degree of polarization was highly correlated between the black and white stripes in each region of interest. While there was a considerable range of d values across zebras and regions of interest, the black and white stripes within each region were similarly matched in their degree of polarization. Black stripes had a greater degree of polarization, however, with a median difference of 46%. Therefore, if tabanids were drawn to polarized light irrespective of its orientation (Egri, Blaho, Sandor, et al. 2012), stripes would confer little benefit to the zebra because they reflect polarized light.

We were interested in what factors might influence the degree of polarization, and the difference between black- and white-stripe d values. The most consistent effect was due to zebra identity (treated as a random factor), which influenced both black- and white-stripe d values. Black-stripe d values were also influenced by whether they were in direct sunlight, but this was the only direct illumination effect; white stripes were not affected by either region or illumination. Figure 5.3 shows (a) the range of d values for black and white stripes from different regions of interest, and (b) when these regions were directly or indirectly illuminated. The regions that systematically showed the greatest differences between black- and white-stripe d values were the flank, back, upper hindleg, rump, and shoulder. Also, black stripes consistently had a higher d value than white stripes, and sunlit regions consistently had a higher d value than shaded regions, as one would expect from optical principles. Nevertheless, none of these differences were substantial (Figure 5.3).

The effect of polarized light on the visual system of a fly approaching from a distance might be influenced not only by the magnitude of the polarization signal, but also by the relative polarization angles in the two sets of stripes. If polarization angles of emitted light were systematically orthogonal between adjacent black and white stripes, then the striping

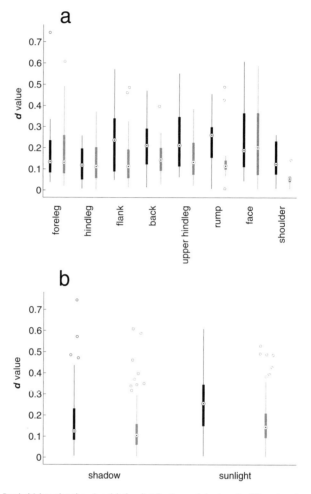

Figure 5.3. Box/whisker plot showing (a) the distributions of *d* values in different regions of interest for black and white stripes and (b) under different illumination conditions. In each column, the dot surrounded by a circle depicts the mean, the box shows the quartiles, the whiskers the 95% confidence intervals, and the "o" symbols show outliers. (*Source:* Britten, Thatcher, and Caro [2016].)

might actively counteract the signal from the black stripes. Therefore, we also analyzed the angle of polarization of the light from our images (see Britten, Thatcher and Caro 2016 for details). Black and white stripes showed similar phase angles, and were highly significantly correlated. Consequently, we concluded that the white stripes cannot counteract the polarized-light signal from the black stripes, because of the consistency of the phase angles of the emitted light from both.

When parasitic flies are tracking hosts, they are thought to shift from olfactory homing to visual homing at relatively short distances of 5–20 m. This behavior is probably a consequence of the low acuity imposed by the limitations of compound eyes. Therefore, we were interested in estimating the distances at which different features of the zebra would be visible to a fly tracking upwind, so we filtered each of our images to emulate a 1° Gaussian blur function at different distances from the zebra. For each of these filtered images, we analyzed the same regions of interest as before. We calculated the d value for the black and white stripes, as well as the luminance contrast between the black and white stripes at several distances out to 50 m (Figure 5.4).

As expected, the Michelson contrast drops steeply with distance, and depending on the size of the stripes, the stripes become invisible at moderate distances of around 10 m. Also, the distinction between the two d values drops at a closer distance, for the same optical reason. But because the d values of the two stripes were similar to begin with (Figure 5.4), they converge to a nonzero value. Thus, at moderate to large distances beyond 20 m, a polarization signal from the whole animal is easier to see than the stripes themselves! Again, this result suggests that stripes are a poor mechanism to defeat polarotaxis, though only if the d value of the zebra coat is higher than that of the background. Thus, we needed to compare the average d value of the zebra and that of the background. We found that coats of the 10 zebras we examined had a significantly higher d value (average of black and white stripe values) than the surrounding vegetation (median ratio 2.06). Therefore, zebras present a distinctive polarization signal, averaged across the black and white stripes, which would be available to a fly approaching upwind and, crucially, at distances where the stripes themselves would be invisible to the fly.

In summary, the main findings to emerge from our study were that substantial polarized light reflectance signals are seen from both black and white stripes in free-living zebras. Second, these are polarized in a similar plane. Third, the polarization of light reflecting from zebras is significantly higher than from surrounding vegetation, so that a positively polarotactic insect would have a distinct signal to guide a visual approach to the zebra. Consequently, it is highly improbable that the primary adaptive significance of the zebra stripes is to defeat polarotaxis by biting flies.

At face value, these results are different from related measurements from domestic ungulates in temperate zones (Horvath et al. 2010; Blaho et al. 2012). The differences are largely in the white stripes, which in our work have a distinct polarization signature. This cannot be attributed to the camera being set at an angle perpendicular to the zebra's flank, since

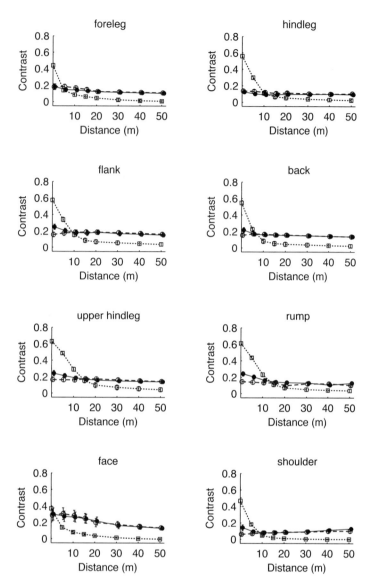

Figure 5.4. Visibility of polarization and stripe contrast with respect to distance from the zebra, plotting each region of interest separately, averaged across individuals. The y-axis represents contrast units, which are *d* values for the polarization data (solid black lines and filled circles denotes black stripes; dashed lines with open circles denote white stripes), and Michelson contrasts (dotted line with open squares) for the luminance data. Both are equivalent, unitless measures of the corresponding optical quantity. Error bars depict SE, and where they are not visible, they are smaller than the plotting symbol. (*Source:* Britten, Thatcher, and Caro [2016].)

many of our regions of interest such as the rump and dorsum were oblique to this angle. Instead, we believe there are at least two other explanations for why differences between the degree of polarization between black and white stripes is modest, and why the angle of polarization is consistent between the two sorts of stripes.

The most striking d values from the Horvath work on live animals were found on dark, dorsal surfaces illuminated by low-angle sun. In the tropics, such illumination conditions are less common, and would mostly occur in the early morning or late afternoon, when tabanids are less actively foraging. Our data included a few observations under such illumination conditions, however, and we observed neither strikingly high black-stripe d values nor strikingly greater differences between black and white stripes across regions (Figure 5.3). This suggests that there are differences between the pelage of zebras in their natural habitat and that of temperate zone ungulates. The pelage of zebras is certainly very thin compared to other artiodactyls (Caro et al. 2014), and likely thinner than temperate zone domestic horses, a factor that would probably increase the relative d values from the white zebra stripes. Thicker pelage would contain a larger number of hairs that light could interact with, which would increase the length and complexity of light paths and thus provide more opportunity for Rayleigh scattering or reflection from randomly oriented surfaces. By contrast, the thin pelage of zebras would encourage a greater fraction of light to be reflected in a systematic way, and thus carry a stronger polarization signal. A second possibility is that a museum zebra pelt used by Horvath, and perhaps his horses, would be less dusty that wild zebras; dust on the hair would both reduce overall polarization and also reduce differences between black and white stripes. Free-living zebras do roll in dust, and dust is always in the air during the dry season in East Africa when our photographs were taken.

Our data provide a challenge to the polarotaxis hypothesis for the adaptive significance of zebra stripes. Polarization signals from wild zebras are largely similar between white and black stripes, in terms of both degree and polarization angle. At the range where flies are thought to transition from olfactory to visual homing (5–10 m), the optics of tabanid eyes would substantially blur the polarization signals of black and white stripes (Figure 5.4), yet the polarization signal of the overall animal would still appear distinct from surrounding dry grass and would thus still be available to attract polarotactic insects. If defeating polarotaxis is not the principal adaptive advantage of stripes, what are the alternative explanations for why biting flies avoid striped surfaces? Parasitic insects typically use a combination of chemical and visual cues to orient to their hosts, with olfaction being more important at longer distances and vision being more important at close range. Tabanids will respond visually to host animals at a distance of about 15 m (Phelps and Vale 1976), and glossinids will abruptly turn toward a visual target from within an odor plume when the target is presented at a distance of 3 m (Torr 1989). Although we know that large, dark-colored objects are maximally effective at attracting temperate zone tabanids, while light-colored, striped, or spotted targets are much less attractive (Allan, Day, and Edman 1987), these conclusions are drawn largely from trapping experiments, leaving open the question of mechanism.

Figure 5.4 shows that the stripes themselves become indistinct at moderate distances where tabanids can visually orient; therefore, the fly must orient to the overall outline of the animal. One possibility is that white stripes may reduce the luminance contrast between the zebra and the background vegetation, compared with a uniformly dark animal, and thus reduce the salience of the zebra as a stimulus for orientation. Another possibility is that the stripes disturb optomotor responses for landing maneuvers. Looming stimuli provoke landing responses if they are in frontal vision, but avoidance turns if presented to the side of a fly (Borst 2014). Patterned stimuli provoke fewer landings relative to uniformly dark ones (glossinids, Turner and Invest 1973; *Drosophila*, van Breugel and Dickinson 2012). While the mechanism for this is not understood, it probably contributes to the protection afforded to zebras by their stripes. These two hypotheses are not exclusive; they operate at different distances. Therefore, stripes may conceal a zebra at moderate to long distances, and disturb fly landing behavior at close range. Interestingly, simulations of motion signals generated by different areas on a zebra's body produce misleading illusions (How and Zanker 2014) that might be similar to that experienced by an incoming fly's own movements.

This chapter has described a large number of studies involving various types of traps and targets, as well as the use of photography, which help to narrow the range of possibilities regarding the effects that striping has on biting flies and the mechanisms by which striping might deter attack. The next step is to refine these methods, especially in regard to trap design and placement, using live animals as targets, and exploring dipteran vision as it relates to striping. I have no doubt that the story will become more complex and interesting with time.

Statistical tests

[1] Spearman rank order correlation coefficients of rates of tail swishing and being bothered, respectively, with outside temperature: zebra, N = 20 sessions, r_s = 0.271, NS; r_s = −0.221, NS; giraffe, N = 20, r_s = 0.043, NS; r_s = 0.471, p = 0.036;

impala, N = 26, r_s = −0.306, NS; r_s = −0.208, NS; warthog, N = 11, r_s = 0.374, NS; r_s = −0.035, NS; topi, N = 11, r_s = −0.497, NS; r_s = 0.521, p = 0.1; buffalo, N = 13, r_s = 0.565, p = 0.044; r_s = −0.249, NS; waterbuck, N = 16, r_s = −0.089, NS; r_s = −0.045, NS.

[2] Spearman rank order correlation coefficients of rates of tail swishing and be-ing bothered, respectively, with windy conditions on a 5-point scale: zebra N = 41 sessions, r_s = −0.258, NS; r_s = −0.051, NS; giraffe, N = 26, r_s = 0.077, NS; r_s = −0.235, NS; impala, N = 38, r_s = −0.102, NS; r_s = −0.227, NS; warthog, N = 17, r_s = −0.371, NS; r_s = −0.061, NS; topi, N = 22, r_s = 0.329, NS; r_s = 0.240, NS; buffalo, N = 13, r_s = 0.086, NS; r_s = −0.244, NS; waterbuck, N = 18, r_s = −0.487, p = 0.041; r_s = −0.039, NS.

[3] Spearman rank order correlation coefficients of rates of tail swishing and being bothered, respectively, with group size: zebra N = 41 sessions, r_s = −0.054, NS; r_s = 0.262, p = 0.097; giraffe, N = 26, r_s = 0.032, NS; r_s = −0.263, NS; impala, N = 38, r_s = 0.092, NS; r_s = 0.097, NS; warthog, N = 17, r_s = −0.308, NS; r_s = −0.199, NS; topi, N = 22, r_s = 0.243, NS; r_s = −0.345, NS; buffalo, N = 13, r_s = 0.304, NS; r_s = 0.106, NS; waterbuck, N = 18, r_s = 0.573, p = 0.013; r_s = 0.377, NS.

[4] Kruskal-Wallis tests of rates of tail swishing and being bothered, respectively, with differing behavior (Ns as above): zebra χ^2 = 0.788, df = 3, NS; χ^2 = 5.563, df = 3, NS; giraffe χ^2 = 6.051, df = 4, NS; χ^2 = 0.378, df = 4, NS; impala χ^2 = 3.062, df = 3, NS; χ^2 = 2.912, df = 3, NS; warthog χ^2 = 2.776, df = 3, NS; χ^2 = 0.727, df = 3, NS; topi χ^2 = 6.924, df = 3, p = 0.074; χ^2 = 1.829, df = 3, NS; buffalo χ^2 = 8.772, df = 4, p = 0.067; χ^2 = 1.357, df = 4, NS; waterbuck χ^2 = 4.813, df = 4, NS; χ^2 = 3.614, df = 4, NS.

[5] Kruskal-Wallis test of rate of tail swishing by species (Ns as above): χ^2 = 86.887, df = 6, p < 0.0001; Kruskal-Wallis test of rate of being bothered by species: χ^2 = 22.349, df = 6, p = 0.001.

[6] Mann-Whitney U tests of rates of tail swishing and rates of being bothered, re-spectively, for zebra vs. giraffe (U = 34.0, p < 0.0001; U = 445.0, NS), zebra vs. impala (U = 95.0, p < 0.0001; U = 644.5, NS); zebra vs. warthog (U = 107.0, p < 0.0001; U = 167.5, p = 0.001); zebra vs. topi (U = 64.0, p < 0.0001; U = 426.0, NS); zebra vs. buffalo (U = 50.0, p < 0.0001; U = 210.5, p = 0.014); and zebra vs. waterbuck (U = 262, p = 0.078; U = 328.5, NS).

[7] t-tests comparing three striped vs. four nonstriped equid species, uncorrected tail swishing, $\bar{X}$s = 8.94, 9.82/individual/min, respectively, t = 0.174, df = 5, NS; uncorrected bothered by flies, $\bar{X}$s = 0.65, 0.72/individual/min, respectively, t = 0.322, df = 5, NS.

[8] Univariate analyses of variance comparing rates of tail swishing by equid species (taking temperature, humidity, and estimated wind speed into account but not reported here). Grevy's zebra vs. plains zebra (df = 1, 23 throughout, F = 1.390, NS); vs. mountain zebra (F = 5.696, p = 0.026); vs. kulan (F = 0.672, NS); vs. Prze-walski's horse (F = 7.729, p = 0.011); vs. African wild ass (F = 11.359, p = 0.003); vs. kiang (F = 2.014, NS). Plains zebra vs. mountain zebra (F = 3.076, p = 0.093); vs. kulan (F = 0.019, NS); vs. Przewalski's horse (F = 4.816, p = 0.039); vs. African

wild ass (F = 20.186, p < 0.0001); vs. kiang (F = 0.481, NS). Mountain zebra vs. kulan (F = 3.498, p = 0.074); vs. Przewalski's horse (F = 0.215, NS); vs. African wild ass (F = 50.341, p < 0.0001); vs. kiang (F = 1.385, NS). Kulan vs. Przewalski's horse (F = 4.131, p = 0.054); vs. African wild ass (F = 24.381, p < 0.0001); vs. kiang (F = 1.202, NS). Przewalski's horse vs. African wild ass (F = 67.016, p < 0.0001); vs. kiang (F = 3.137, p = 0.090). African wild ass vs. kiang (F = 31.073, p < 0.0001).

[9]   Univariate analyses of variance comparing rates of being bothered by flies by equid species (taking temperature, humidity, and estimated wind speed into account but not reported here). Grevy's zebra vs. plains zebra (df = 1, 23 throughout, F = 4.313, p = 0.049); vs. mountain zebra (F = 0.572, NS); vs. kulan (F = 2.010, NS); vs. Przewalski's horse (F = 0.731, NS); vs. African wild ass (F = 6.345, p = 0.019); vs. kiang (F = 2.885, NS). Plains zebra vs. mountain zebra (F = 13.171, p = 0.001); vs. kulan (F = 1.032, NS); vs. Przewalski's horse (F = 2.810, NS); vs. African wild ass (F = 0.070, NS); vs. kiang (F = 0.912, NS). Mountain zebra vs. kulan (F = 11.475, p = 0.003); vs. Przewalski's horse (F = 5.538, p = 0.028); vs. African wild ass (F = 27.403, p < 0.0001); vs. kiang (F = 13.415, p = 0.001). Kulan vs. Przewalski's horse (F = 0.384, NS); vs. African wild ass (F = 2.496, NS); vs. kiang (F = 0.065, NS). Przewalski's horse vs. African wild ass (F = 3.525, p = 0.073); vs. kiang (F = 1.336, NS). African wild ass vs. kiang (F = 0.419, NS).

[10]   Mann-Whitney U tests comparing three striped vs. four nonstriped equid species, corrected tail swishing, $\bar{X}s$ = 3.16, 4.74/individual/min/flying insect × 100, respectively, U = 6, NS; corrected bothered by flies, $\bar{X}s$ = 0.20, 0.31/individual/min/fly × 100, respectively, U = 5, NS.

[11]   Plains zebra in Tierpark zoo: Spearman rank order correlation coefficient between rates of tail swishing and the proportion of black pelage on the rump (N = 5 individuals, r_s = –0.900, p = 0.037).

[12]   Grevy's zebra in Tierpark zoo: Spearman rank order correlation coefficients between rates of tail swishing and average lateral black stripe width (N = 5 individuals, r_s = 0.957, p = 0.010), and between rates of being bothered by flies and percentage of the neck with black stripes (N = 5, r_s = –0.866, p = 0.058).

[13]   Mountain zebra in Tierpark zoo: Spearman rank order correlation coefficient between rates of tail swishing and the ratio of widest black to widest white stripe (N = 7 individuals, r_s = 0.739, p = 0.058).

[14]   Three zebra species in Tierpark zoo combined: Spearman rank order correlation coefficients between rates of tail swishing and rates of being bothered, each standardized by insect (fly and hymenopteran) abundance with the ratio of widest black to widest white stripe (N = 17 individuals, r_s = 0.608, p = 0.010; r_s = 0.774, p < 0.0001, respectively). Spearman rank order correlation coefficients between rates of tail swishing and rates of being bothered, each standardized by insect abundance and average lateral black stripe width (N = 17 individuals, r_s = 0.422, p = 0.092; r_s = 0.428, p = 0.087, respectively).

[15]   Friedman test across 6 sets of biconical trap sessions for Glossinidae χ^2 = 11.385, df = 4, p = 0.023; Wilcoxon tests, blue vs. dark brown ($\bar{X}s$ = 6.33, 1.17, respectively, z = 2.023, p = 0.043); blue vs. light brown/black ($\bar{X}s$ = 6.33, 1.67, z = 2.023, p =

0.043); dark brown vs. light brown ($\bar{X}$s = 2.67, 1.17, z = 2.264, p = 0.024); all other paired comparisons NS.

[16] Friedman tests across 6 sets of biconical trap sessions for Tabanidae χ^2 = 4.000, df = 4, NS; for other arthropods χ^2 = 8.108, df = 4, p = 0.088; Wilcoxon test, light brown/black vs. striped ($\bar{X}$s = 72.67, 85.50, respectively, z = –1.761, p = 0.078; all other paired comparisons NS).

[17] Friedman test across 6 sets of cloth trap sessions for Glossinidae χ^2 = 16.100, df = 4, p = 0.003; Wilcoxon tests, black vs. white ($\bar{X}$s = 2.67, 0.33, respectively, z = 2.060, p = 0.039); black vs. horizontal striped ($\bar{X}$s = 2.67, 0, z = 2.060, p = 0.039); black vs. vertical striped ($\bar{X}$s = 2.67, 0.50, z = 1.841, p = 0.066); gray vs. white ($\bar{X}$s = 1.50, 0.33, z = 1.890, p = 0.059); gray vs. horizontal striped ($\bar{X}$s = 1.50, 0, z = 2.060, p = 0.039); gray vs. vertical striped ($\bar{X}$s = 1.50, 0.50, z = 1.890, p = 0.059); all other paired comparisons NS.

[18] Friedman test across 6 sets of cloth trap sessions for Tabanidae χ^2 = 13.892, df = 4, p = 0.008; Wilcoxon tests, black vs. white ($\bar{X}$s = 2.83, 0.50, respectively, z = 2.032, p = 0.042); black vs. horizontal striped ($\bar{X}$s = 2.83, 0, z = 2.032, p = 0.042); black vs. vertical striped ($\bar{X}$s = 2.83, 0.17, z = 2.032, p = 0.042); white vs. horizontal striped ($\bar{X}$s = 0.50, 0, z = 1.732, p = 0.083); all other paired comparisons NS.

[19] Friedman test across 6 sets of cloth trap sessions for arthropods χ^2 = 12.739, df = 4, p = 0.013; Wilcoxon tests, black vs. gray ($\bar{X}$s = 286.5, 79.5, respectively), z = 2.201, p = 0.028; black vs. white ($\bar{X}$s = 286.5, 74.67), z = 2.201, p = 0.028; black vs. horizontal striped ($\bar{X}$s = 286.5, 85.83), z = 2.201, p = 0.028; black vs. vertical striped ($\bar{X}$s = 286.5, 117.83), z = 1.992, p = 0.046); all other paired comparisons NS.

[20] Wilcoxon matched pairs tests of black balls and horizontally striped black and white balls in canopy traps, N = 11 pairs of traps, tabanids (*Tabanus* and *Atylotus* combined) z = –2.371, p = 0.018; glossinids, z = –1.890, p = 0.059.

[21] Univariate analyses of variance comparing all tabanids captured over 48 trap days in zebra and wildebeest canopy traps, taking folding on the western aspect and total trap folds into account. Model F = 6.096, df = 3, 92, p = 0.001; west side folding F = 10.869, df = 1, 92, p = 0.001; total folds F = 7.403, df = 1, 92, p = 0.008; pelt type F = 3.883, df = 1, 92, p = 0.052.

[22] Univariate analyses of variance comparing live tabanids captured over 48 trap days in zebra and wildebeest canopy traps, taking folding on the western aspect and total trap folds into account. Model F = 5.574, df = 3, 92, p = 0.001; west side folding F = 4.778, df = 1, 92, p = 0.031; total folds F = 7.583, df = 1, 92, p = 0.007; pelt type F = 6.949, df = 1, 92, p = 0.010.

Univariate analyses of variance comparing dead tabanids captured over 48 trap days in zebra and wildebeest canopy traps, taking folding on the western aspect and total trap folds into account. Model F = 4.6816, df = 3, 92, p = 0.004; west side folding F = 9.596, df = 1, 92, p = 0.003; total folds F = 5.273, df = 1, 92, p = 0.024; pelt type F = 2.212, df = 1, 92, NS.

[23] Univariate analyses of variance comparing nontabanid dipterans captured over 48 trap days in zebra and wildebeest canopy traps, taking folding on the western

aspect and total trap folds into account. Model F = 4.420, df = 3, 92, p = 0.006; west side folding F = 6.1741, df = 1, 92, p = 0.015; total folds F = 5.374, df = 1, 92, p = 0.023; pelt type F = 4.209, df = 1, 92, p = 0.043.

[24] Spearman rank order correlations of the proportion of canopy pelt trap that was of black pelage and numbers of tabanids captured (N = 6 balls, r_s = 0.115, NS), and the frequency of black stripes on zebra pelt canopy traps and numbers of tabanids captured (N = 6 balls, r_s = 0.572, NS).

[25] Friedman test across 25 matched 600 m walks in suits for Glossinidae χ^2 = 30.818, df = 4, p < 0.0001; Wilcoxon tests, black vs. gray (z = 3.632, p < 0.0001; black vs. white (z = 2.252, p = 0.024); black vs. vertical striped (z = 2.327, p = 0.020); gray vs. white (z = −2.169, p = 0.030); gray vs. horizontal striped (z = −3.840, p < 0.0001); white vs. horizontal striped (z = −2.456, p = 0.014); horizontal striped vs. vertical striped (z = 2.321, p = 0.020); all other paired comparisons NS.

[26] Friedman test across 25 matched 600 m walks in suits for insects χ^2 = 31.980, df = 4, p < 0.0001; Wilcoxon tests, black vs. gray (z = 3.727, p < 0.0001); gray vs. white (z = −3.100, p = 0.002); gray vs. horizontal striped (z = −4.117, p < 0.0001); gray vs. vertical striped (z = −2.408, p = 0.016); white vs. horizontal striped (z = −1.933, p = 0.053); horizontal striped vs. vertical striped (z = 2.271, p = 0.023); all other paired comparisons NS.

[27] Wilcoxon test on the number of glossinids landing on me or the pelt when I was wearing the zebra or wildebeest pelt (37 paired walks, $\bar{X}$s = 6.35, 7.57, respectively), z = −1.541, NS.

[28] Wilcoxon test on the number of tsetse flies landing on black or white stripes of the zebra pelt following 37 paired walks ($\bar{X}$s = 2.78, 3.03, respectively), z = −0.383, NS.

[29] Wilcoxon test on the number of insects landing on me or the pelt when I was wearing the zebra or wildebeest pelt (N = 37 paired walks, $\bar{X}$s = 7.38, 8.92), z = −1.906, p = 0.057.

[30] Wilcoxon tests on the number of glossinids and insects landing on the zebra pelt or wildebeest pelt on 25 slow drives. Glossinids ($\bar{X}$s = 7.84, 4.08, respectively), z = 3.349, p = 0.001. Insects ($\bar{X}$s = 9.18, 4.94, respectively), z = 2.733, p = 0.006.

[31] Wilcoxon test on the number of tsetse flies landing on black or white stripes of the zebra pelt following 25 slow drives ($\bar{X}$s = 4.13, 4.04, respectively), z = −0.131, NS.

Figure 6.0. Mutual grooming is thought by some to be promoted by the presence of pelage stripes. (Drawing by Sheila Girling.)

SIX

Intraspecific communication

6.1 Intraspecific signaling

Because people can see zebras so easily at close range and can instantly recognize them as zebras, some biologists have proposed that stripes are concerned with recognition or with facilitating social behavior. There are several related but separate issues under this general hypothesis (reviewed in Ruxton 2002). The first concerns species recognition: here stripes might help zebras distinguish members of their own species from heterospecifics, perhaps from other zebra species in areas of sympatry (geographic range overlap) (Morris 1990). Second, Kingdon (1979, 1984) proposed that stripes facilitate social grooming. He thought that the neck and withers regions of the body are preferred target areas of mutual grooming in horses (Feh and de Mazieres 1993) and also in zebras and suggested that stripes could have evolved to mimic skin wrinkling in these areas seen when an equid turns its head. Under this hypothesis, striping advertises these areas and so promotes allogrooming (mutual grooming of conspecifics), and additionally promotes social bonding. Cloudsley-Thompson (1984) concurred, suggesting that indeed there are specific areas of the body that are presented to conspecifics for grooming and that body stripes call attention to these areas, and thereby reinforce bonding between mother and foal, and between mare and stallion.

Another possibility is that striping facilitates individual recognition (Morris 1990); individuals might learn and recognize the striping patterns of others. For example, Wallace (1877) thought that "the stripes therefore may be of use by enabling stragglers to distinguish their fellows at a distance" (p. 400). A fourth hypothesis is that aspects of striping are signals of quality involved in mate choice or in contest competition among males; the idea of striping signaling quality has been put forward in relation to escape potential from predators (Ljetoff et al. 2007) but not in relation to sexual selection. Although these hypotheses can be cursorily dismissed on logical grounds that unstriped equids groom one another, recognize one another, and fight and reproduce without the putative social benefits of striping, the ideas still need to be tested.

6.2 Species recognition

The species recognition argument suffers from logical difficulties. First, most ungulate species appear to recognize conspecifics in that they associate with members of their own species without the need for striking markings. Unstriped equids do this too. Only in Neotropical primates (Santana, Alfaro, and Alfaro 2012) and African Cercopithecines (Santana et al. 2013; Allen, Stevens, and Higham 2014) has pelage coloration in mammals has been linked to species recognition (and reduction in hybridization between species). In these groups of primates, species in greater sympatry have more complex facial markings. These taxonomic groups consist of closely related species that have considerable geographic range overlap, are highly social, and in many cases have trichromatic vision.

In contrast, zebras show little sympatry with other equids. Only Grevy's and plains zebra, and mountain and plains zebra show geographic range overlap and this in a small proportion of their historical ranges. Across a large proportion of their geographic range, plains zebras are allopatric with other equids (i.e., don't overlap with other members of the genus). In addition, surprising differences in equid chromosome numbers mean that cross-species mating between equid species will not easily result in viable offspring, and appear to occur only in exceptional circumstances (Schieltz and Rubenstein 2015).

6.3. Stripes as a facilitator of mutual grooming and social bonding

Kingdon (1979) developed his ideas about an association between striping and grooming thus: "It is common for a preferred grooming area to be differentiated in the texture, colour or pattern of the fur. When an equid raises its head, the skin at the base of the neck wrinkles. In a short-maned, short-haired horse, this folding of the skin is perceptible to the eye and to the touch. Because this is also a target area for grooming, it is conceivable that fine barring on the withers might have originated as an enhancement or mimicry of this physical characteristic to guide a partner to the preferred site for nibbling" (p. 135). Kingdon next went on to suggest that grooming could facilitate socialization, and he observed that members of a herd of Grevy's zebras come to a halt parallel to each other to a far greater extent than do domestic horses after running away from a source of danger, and remarked that shorter nearest neighbor distances might have some connection to the narrow frequency of stripes in this species. He exposed captive Grevy's zebras to rectangular panels painted with stripes of differing width and observed that they chose to stand closer to panels with thinner stripes. Finally, he wrote that "once the stripes actually played a positive role in promoting sociability they would have lost their subsidiary function as foci for grooming and their effectiveness would have been diminished by being limited to certain parts of the body so that a more efficient use of the mechanism required the extension of striping to embrace the whole animal. Since the stripes may have their optimal effect when individuals are at a specific distance from each other, the stripes might have acquired the secondary function of helping, in certain circumstances, to space zebras" (p. 137). Kingdon (1979) thought that other equid species living in dispersed units at low densities in harsher environments would no longer need striping to facilitate the social functions that he proposed, and additionally noted that in species that acquired thick winter coats, fleeces would obscure any crisp striping patterns.

There are several facets of Kingdon's argument: a mechanistic evolutionary scenario for striping being a form of intraspecific communication; linking grooming to the maintenance of both social relationships and spatial distributions in equids; and an explanation for why only three of the seven equid species are striped. These ideas are difficult to test because the first suggests that grooming might no longer (in evolutionary terms) be directed at the withers, neck, and mane but all over the body; and that striping might serve to keep individuals close together or far apart. Also, he provides two explanations for lack of striping: living in desert environments at low densities, and molting, which are difficult to test with only seven species of equids.

6.3.a Allogrooming

Grooming between conspecific mammals is known to be influenced by parasite load, weather, seasonal effects, and familiarity (R. Barton 1985; Schino et al. 1988). If facilitation of grooming by other animals is the evolutionary driver of striping patterns in zebras, zebras should groom one another frequently. To examine this, I watched the behavior of groups of 3–15 plains zebras (297 total animals in Katavi) for a total of 50.25 hours across 38 sessions on open plains during daylight hours. Zebras stayed in close proximity to one another, an average of 3.1 estimated body lengths (SE = 0.7), with the group spread out over an estimated 50.1 m (SE = 18.4), so there was ample opportunity to groom other animals. Indeed, the percentage of time that an individual rested with its lower mandible leaning on another animal's shoulder was substantial, 4.10% (SE = 1.17, N = 33 sessions), so individuals could easily begin to groom one another. Yet the rate at which plains zebras actually groomed one another was very low indeed (0.02/hr, SE = 0.008, N = 38 sessions), and the percentage of time that an individual zebra groomed another was again low: 0.15% (SE = 0.07, N = 38 sessions) of the time. The percentage of time that an individual groomed itself was also very low: 0.64% (SE = 0.14, N = 38 sessions). Others have reported higher rates of both activities (Chapter 1 and Table 6.1), and I am not sure why Katavi plains zebras groomed so little.

Allogrooming occurs more frequently among unstriped domestic horses. For instance, allogrooming takes place frequently in Icelandic horses, at 0.24/hr, especially between individuals of the same age and sex (Sigurjonsdottir et al. 2003), and at 0.65/hr in Scottish highland ponies (Clutton-Brock, Greenwood, and Powell 1976), although less often between Camargue horses (0.06–0.13/hr) (Wells and Goldschmidt-Rothschild 1979). Limited data on mutual grooming across wild species of equid indicate that some striped and unstriped species allogroom a little, but some do so a lot (Table 6.1). Grevy's zebras and African wild asses groom one another relatively infrequently, whereas plains zebras, mountain zebras, and Przewalski's horses groom one another far more often for relatively long periods of time during the day.

Table 6.2 shows that most bouts of allogrooming performed by plains zebras on floodplains in Katavi in the dry season were directed toward the rump, followed by the neck and then the midflank region (Plate 6.1). Time spent grooming the rump ($\bar{X}$ = 14.5 s) and neck ($\bar{X}$ = 5.4 s) was far higher than any other area ($\bar{X}$s of all < 1 s). Note that these areas are not distinguished from other parts of the body in having thin vertical stripes (Table

Table 6.1. Qualitative descriptions of mutual grooming (allogrooming) in equids.

Plains zebra	Grooming, especially neck/shoulders. Mare/foal and stallion/mare were preferred grooming partners, but mare/mare was not observed.[a]
	Mare/foal, not mare/mare. As much as ½ hour, several times a day. Grooming was sustained longest after a parting.[b]
	Males take part in mutual grooming between 0.2 and 0.3 times a day on average; females participate in allogrooming around 0.1 times a day on average.[c]
Mountain zebra	Grooming most frequent between mare/foal; may occur between mare/stallion, foal/stallion, and mare/mare as well. Individual grooming by shaking, nibbling, scratching, dust bathing at least once per day.[d]
	Grooming maintains group cohesion. Mare/foal most common.[e]
	Mutual grooming occurs for up to 20 min at a time.[f]
Grevy's zebra	Allogrooming very rare among individuals.[g]
	Very low rates of grooming.[c]
African wild ass	Adult males had no specific grooming partners, groomed with other males. Females groom less often, usually female/female or female/foal. Adult females and adult males were not observed mutual grooming.[h]
Asiatic wild ass	Common.[i]
Kiang	Uncommon.[j]
Przewalski's horse	Closely related mares prefer one another as grooming partners. Colts spent as much time grooming with stallions as expected, while fillies spent much less time than expected.[k]
	Male 2–3-year-olds deepen personal contacts in their rank group by grooming, games, and scuffling.[l]
	Self-groom: 1.9 ± 0.3% of time budget; mutual groom: 1.1 ± 0.1% of time budget.[m]
	Mutual grooming makes up 11.2 and 4.3% of the time budget for nonagonistic acts for Przewalski's horses at two different study sites.[n]

[a]Klingel (2013).
[b]Kingdon (1979).
[c]Kimura (2000).
[d]Penzhorn (1988).
[e]Penzhorn (2013).
[f]Skinner and Chimimba (2005).
[g]S. Williams (2013).

[h]Moehlman (1998b).
[i]Prothero and Schoch (2003).
[j]Xu et al. (2013).
[k]Boyd and Houpt (1994).
[l]Klimov (1988).
[m]Boyd (1991).
[n]Keiper and Receveur (1992).

Table 6.2. Allogrooming by plains zebras in Katavi.

Area of body	Proportion of grooming bouts	Mean duration	Stripe thickness	Stripe alignment
Rump	32.5	14.5	Wide	Horizontal
Neck	28.3	5.4	Medium	Oblique
Midflank	15.5	0.6	Wide	Vertical
Proximal flank	8.8	0.2	Medium	Vertical
Dorsum	7.7	0.2	Medium	Horizontal
Distal flank	3.6	0.3	Wide	Oblique
Head	1.8	0.1	Thin	Oblique
Hindleg	1.3	0.1	Thin	Horizontal
Chest	0.6	0.3	Wide	Oblique
Ventrum	0	0	Medium	Vertical

Notes: Proportion of bouts of grooming other zebras (allogrooming) that were directed to different areas of the conspecific's body (N = 297 zebras) and average durations of grooming bouts in seconds. Allogrooming was defined as nibbling with teeth or head rubbing areas of another zebra's body.

Table 6.3. Mean (and SD) percentage of body length (nose-anus) that can be reached by the tail, and percentage of tail length that is covered by tail hairs for zebras, other equids, and selected African artiodactyls taken from museum specimens.

Species	N	Percentage of body reached by tail	Percentage of tail length with tail hairs
Plains zebra	13	47.8 (4.2)	65.4 (2.2)
Mountain zebra	2	45.5 (3.1)	59.8 (1.3)
Grevy's zebra	8	44.2 (4.8)	58.8 (1.9)
Asiatic wild ass	7	49.6 (9.3)	66.3 (2.3)
Przewalski's horse	2	46.6 (5.9)	69.7 (3.5)
Blue wildebeest	8	61.7 (13.3)	66.2 (3.8)
Roan antelope	10	46.2 (8.4)	59.3 (3.7)
Beisa oryx	9	47.2 (10.5)	64.3 (5.4)
Hartebeest	7	39.5 (3.4)	58.7 (2.6)
Waterbuck	8	28.6 (4.2)	59.4 (1.9)
Topi	12	41.9 (4.1)	59.2 (3.2)

Source: H. Walker, unpublished data.

6.2), as Kingdon's wrinkle mimicry idea might predict. Qualitative data presented here suggest no bias in allogrooming toward certain thicknesses or orientation of stripes. It is interesting that allogrooming was predominantly directed toward the rump area because the idea that stripes promote social grooming grew out of stripes mimicking wrinkled skin on the withers, neck, and mane, where the subject cannot reach to groom itself. If striping then spread over the body during the course of evolution to attract groomers' gaze to other areas that need attention, it is strange that conspecifics concentrate their grooming activities on the rump, which is easily reached by the tail and is therefore more likely to be free of ectoparasites (Siegfried 1990) (Table 6.3). Equids and African herbivores show similar proportions of the body that can be reached by the tail and similar proportions of the tail that are covered by tail hairs that splay out and flay a larger area of the body than a fleshy tail tip [1, 2]. Given that zebras do not appear morphologically handicapped in being able to remove flies by means of tail swishing (Table 6.3), it seems odd that they need stripes to attract allogrooming to the rump. It is the lack of association between striped equids and extensive mutual grooming that is most damaging to the allogrooming hypothesis, however.

6.3.b Social bonding

Kingdon (1979) linked pelage stripes to allogrooming and then to the formation or maintenance of social bonds. Aside from removing ectoparasites

(Tyler 1972), grooming certainly has a social function in equids (see Crowell-Davis, Houpt, and Carini 1986). As an illustration, grooming directed at the withers reduces heart rate in Camargue horses (Feh and de Mazieres 1993). Nonetheless, there are a number of observations that speak against striping being involved in forming or maintaining social relationships. First, data collected on free-ranging horses on Yururi Island, Japan, show that preferred allogrooming partners are not the same individuals as the nearest neighbors with whom individuals associate (Kimura 1998). Second, if striping promotes social bonding, we would expect to find Type I equids that live in permanent small harems to be striped, as is the case in plains and mountain zebras. But Type I equids also include the uniformly brown Przewalski's horse as well as free-ranging domestic horses that are completely uniform or sport large blocks of white, brown, or black pelage. Grevy's zebras, which live in widely dispersed groups, are a Type II equid. Third, striping is not associated with living at high densities, where the necessity to form social bonds might be paramount. Table 1.3 shows that kiang and Asiatic wild asses aggregate in large numbers, whereas Grevy's zebras live at low densities, both of which are counterexamples to this proposal. Fourth, anecdotal reports of rare-colored individual zebras provide mixed support for these being avoided by conspecifics. Kingdon (1984) wrote, "When a black zebra was seen, in the Rukwa Valley, Tanzania, it tended towards a peripheral position whenever its group was buzzed by aircraft or approached by a vehicle" (p. 487), whereas Matthews (1971) remarked, "An unusual colour anomaly has been seen in zebra in Rhodesia (now Zimbabwe) in recent years—an animal is black with white spots, the black stripes having encroached upon the white ones. This extraordinary individual—which looks like a child's toy—has been photographed from the air and does not appear to be an outcast from the herds of its normally coloured relatives or in any way handicapped in earning its living" (p. 350). Finally, in regard to molt being unable to provide crisp stripes for attracting grooming, absence of stripes is, as Kingdon surmised, found in the three Asiatic equid species, all of which undergo annual molts, yet the African wild ass, largely lacking stripes, has no pronounced molt. We can summarize these disparate pieces of information by saying that there is no support whatsoever for predictions of the social bonding and striping hypothesis (Caro in press).

6.4. Stripes as a means of individual recognition

Domestic horses are dichromats (Geisbauer et al. 2004; Hanggi, Ingersoll, and Waggoner 2007), but their vision is surprisingly similar to that of humans because their large eye gives them sensitivity in achromatic tasks (Roth, Balkenius, and Kelber 2008). They can discern stripes of most species at 75 m or more under photopic and mesopic conditions (Table 2.12), better than predators (Table 2.11). Humans can identify individual zebras by their differing striping patterns, leading to suggestions that the function of striping is individual recognition by conspecifics (Morris 1990). For example, from photographs I could tell individuals apart by the unique pattern of markings at the top of the foreleg where horizontal stripes formed triangular arrangements with vertical body stripes, by attending to y-shaped splits in black or white midflank stripes that divide into vertical body stripes and oblique rump stripes, by noting partially split black or white stripes on the neck or flank, by examining curlicue stripes around the groin and upper hindleg, and by finding black broad stripes that ran forward from the rear portion of the rump but abruptly stopped in the distal (rear) region of the flank (see also Ginsberg and Rubenstein 1990).

Computer programs can also identify individual zebras (Foster, Krijger, and Bangay 2007). Nonetheless, in nature discriminations require time and might necessitate zebras standing still, making details of striping difficult to utilize in the field. Instead, coarse aspects of striping might be more practical for zebras. I therefore analyzed photographs of plains zebras taken in Katavi using Photoshop, as described in section 4.7.b, and subsequently derived measures of stripe number, stripe thickness, and proportions of different parts of the body that are black. These coarse quantitative measures of stripe patterning confirmed that individuals differ from one another in appearance. Taking those 151 photographs of 61 individuals for which I had more than one photograph (taken days, weeks, or years apart), I found that photographs of the same individuals differed significantly less from each other than photographs of different individuals on the following measures: total number of stripes on neck and flank, the average stripe width on this part of the body, the ratio of widest black to widest white stripe, and the percentage of black pelage on flank and rump, although not on the neck [3]. All of these measures except the amount of black on the neck could theoretically be used to distinguish individuals.

So coarse aspects of striping or individual stripe patterns themselves could be used as aids in individual recognition, but are they used in this way? There is no strong evidence for this. First, striping is not restricted to one of the two types of equid social organization—neither to Type I, where harem members are in constant association, nor to Type II, where widely dispersed females and males might require help in recognizing one another after being apart for protracted periods of time. Second, pelage

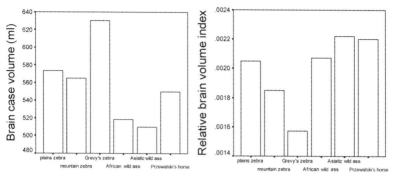

Figure 6.1. Mean brain case volumes of plains zebra (N = 10), mountain zebra (N = 1), Grevy's zebra (N = 11), African wild ass (N = 4), Asiatic wild ass (N = 8), and Przewalski's horse (N = 3), and relative brain volume index (volume / body weight). (H. Walker, unpublished data.)

striping is not restricted to species with smaller brain sizes that might be less capable of identifying other individuals of their own species and so require additional aids to conspecific recognition. Neither is striping found in species with larger brain sizes that might be capable of more complex discrimination tasks. Absolute brain volume is greater in striped than unstriped equids; relative brain size to body weight is lower than or similar to that of their unstriped congeners (Figure 6.1). In either case, a single badge of contrasting coloration might better identify an individual than stripes.

Third, equids are a very vocal genus (Policht, Karadzos, and Frynta 2011; Rubenstein 2011), with a large repertoire that carries information about social status (Rubenstein and Hack 1992), familiarity, sex, body size (Basile et al. 2009; Lemasson et al. 2009), and individual identity (Tyler 1972; Wolski, Houpt, and Anderson 1980; Lemasson et al. 2009; Proops, McComb, and Reby 2009). The complexity of information conveyed by the auditory modality casts aspersions on the idea of additionally needing visual signals for individual recognition. Also, dung odor conveys information about familiarity in feral horses without recourse to visual signals (Rubenstein and Hack 1992).

Fourth, domestic horses are highly intelligent and are capable of sophisticated individual recognition employing both auditory and visual information (without stripes). Examining two herds of horses of around 30 animals each, Proops and colleagues (2009) found that horses show cross-modal individual recognition of herd members. They let horses see another familiar individual being led past them and then disappear behind a barrier and then, more than 10 sec later, played a call of the same or of a different individual from behind the partition. Subjects looked significantly longer and responded more quickly when the auditory stimulus came from an animal different from the one they had just witnessed disappear. Such social knowledge extended to all horses tested and was no greater in older animals or those of a particular sex. Given that cross-modal individual recognition (here auditory and visual) occurs in a non-striped equid, one could argue either that zebras were once incapable of using uniform pelage for mutual recognition and needed stripes or, more parsimoniously, that striping has little to do with individual recognition.

6.5. Stripes as an indicator of quality

A vaguely formulated argument for the evolution of striping in equids is that aspects of striping are an indicator of genetic quality that predators might use in making hunting decisions (see section 4.8). The proposal can be extended to striping being a signal to other zebras that is used in mate choice or in contests over territory or mates, or perhaps to settle contests over limited resources such as access to a waterhole. For example, individual zebras show huge variation in the amount of black on their pelage (Table 6.4, Plate 6.2), and in some mammals dark pelage is associated with age or dominance. This occurs in males of some artiodactyl species (Caro 2011) and in lions, where dark manes are associated with ability to win fights and attract females (P. West and Packer 2002). In equids striping patterns do not change once infants lose their brown coat, except that stripes become broader as the animal grows in size (Murray 2007), so stripe number and later stripe width are unlikely to be a signal of short-term changes in condition but could nevertheless indicate something about early development in utero, perhaps concerning genetic characteristics. Measures of genetic quality cannot be assessed easily under field conditions, but a limited set of proxies are available and I examined some of these here.

Table 6.4. Percentages of different areas of the body of 287 plains zebras in Katavi National Park that are black.

	Mean	SD	Min	Max
Percentage of neck that is black	57.8	8.7	10.6	74.2
Percentage of flank that is black	47.9	7.2	12.6	98.4
Percentage of rump that is black	55.7	6.4	22.8	67.1
Total number of stripes on neck and flank	23.1	1.8	17	28
Ratio of widest black stripe to widest white stripe	0.60	0.06	0.16	0.73

Notes: Means, standard deviations, minima, and maxima are given. Also shown is the ratio of black to white stripes.

If striping patterns were sexually selected one might expect to see sex differences in patterning but almost none were evident in the quantitative measures that I obtained of Katavi plains zebra using Photoshop when I considered subadults and adults together [4] or simply adult individuals [5]. Adult males had marginally fewer lateral stripes than adult females. That said, in the field I was convinced that adult males were slightly shorter at the shoulder than adult females, that they had a greater expanse of black pelage, and that they had thinner white stripes especially on the rump. This was not supported by the data, however [6]. Perhaps I was being influenced by males having blacker stripes than females? In plains zebras living in western Tanzania, then, the extent of striping does not appear to be a sexually selected trait suggesting it is not used to settle contests or in female choice.

If striping patterns are an indicator of genetic quality, they might be reflected in phenotypic traits. I therefore tested whether striping measures were associated with calculated shoulder heights and body weights, since the former and to some extent the latter are set during development by both environmental and genetic factors. There was little evidence that any of the six striping measures were related to body size save that the percentage of black pelage on the rump and perhaps average stripe width were significantly positively correlated with calculated weights [7]. The amount of black on the rump is therefore a cue to body weight insofar as my estimates are correct.

Fluctuating asymmetry in morphometric characters has been used as a measure of developmental stability and by extension genetic quality. This is because better complements of genes are thought to be able to buffer environmental perturbations at the cellular level during development. The logic of this argument is no longer as strong as it used to be, but I thought I would examine it cursorily in free-living zebras. I simply scored by eye each individual as having symmetrical or asymmetrical striping patterns when viewed directly from the front or from the rear, areas of the body not used in quantitatively assessing striping patterns. There were no associations between any of the six striping measures and these two measures of stripe symmetry [8, 9]. In conclusion, none of these four sets of observations help make a case for stripes being an indicator of phenotypic quality.

Finally, I examined whether individuals with particular patterns of striping showed greater evidence of wounding—namely, open wounds, fresh or old scars, or missing tails—since predators might target low-quality individuals or those in poor condition on the basis of their stripes. I found no difference in striping patterns between individuals that showed signs of wounding and those that did not [10].

6.6 Conclusion

The hypothesis that stripes serve a social function is intriguing but difficult to test. Where I have been able to tease out predictions, I find no support for any variant of this hypothesis. Stripes promoting species recognition certainly seem improbable given the limited extent of sympatry in the three species of zebra. Following on from that, none of the observations presented in this chapter support any aspect of striping being related to allogrooming, social bonding, or individual recognition or being an indicator of quality. This is perhaps unsurprising because other equids and large mammalian herbivores live in social groups where they allogroom, bond, and recognize one another without recourse to striped pelage. In regard to allogrooming, qualitative descriptions indicate that it is relatively common among unstriped equids, so striping cannot be a prerequisite for this behavior, while low rates of allogrooming in some zebra populations suggest that striping is unlikely to be associated with conspecific grooming activity. Neither is striping related to crude categories of social organization—namely, harem defense polygyny (Type I) or resource defense (Type II). Domestic horses are capable of sophisticated individual recognition using visual cues in the absence of stripes, and it is stretching imagination to argue that closely related zebras need stripes to do this. Finally, no data lend support to striping being associated with phenotypic quality.

Statistical tests
[1]   t-test comparing head to anus length divided by tail length across equids (i.e., plains zebra [N = 13], mountain zebra [N = 2], Grevy's zebra [N = 8], Asiatic wild ass [N = 7], and Przewalski's horse [N = 2] together) and artiodactyls (i.e., blue wildebeest [N = 8], roan antelope [N = 10], beisa oryx [N = 9], hartebeest [N = 7], waterbuck [N = 8], and topi [N = 12] together), t = 1.224, NS.
[2]   t-test comparing tail hair length divided by tail length across equids (i.e., plains zebra [N = 13], mountain zebra [N = 2], Grevy's zebra [N = 8], Asiatic wild ass [N = 7], and Przewalski's horse [N = 2] together) and artiodactyls (i.e., blue wildebeest [N = 8], roan antelope [N = 10], beisa oryx [N = 9], hartebeest [N = 7], waterbuck [N = 8], and topi [N = 12] together), t = 2.899, NS.
[3]   One-way ANOVAS comparing striping patterns within and between 61 individuals based on 151 photographs, total number of neck and flank stripes, F = 2.231, df = 61, 90 in all cases, p < 0.0001; average stripe width on neck and flank, F = 1.974, p = 0.002; ratio of widest black and white stripes, F = 5.160, p < 0.0001; percentage of neck black, F = 1.326, NS; percentage of flank black, F = 3.254, p < 0.0001; percentage of rump black, F = 1.935, p = 0.002.

[4] t-tests comparing striping patterns of subadult and adult male and female zebras (N = 59, 186, respectively), total number of neck and flank stripes, $\bar{X}$s males and females, respectively = 22.8, 23.1, t = −1.142, df = 243, NS; average stripe width on neck and flank, $\bar{X}$s = 56.3, 55.4 pixels, t = 0.394, df = 243, NS; ratio of widest black and white stripes, $\bar{X}$s = 0.61, 0.60, t = 1.566, df = 243, NS; percentage of neck black, $\bar{X}$s = 57.5%, 57.8%, t = −0.231, df = 75.92, NS; percentage of flank black, $\bar{X}$s = 47.9%, 48.6%, t = −0.545, df = 243, NS; percentage of rump black, $\bar{X}$s = 55.9%. 55.7%, t = 0.178, df = 243, NS.

[5] t-tests comparing striping patterns of adult male and female zebras (N = 48, 152, respectively), total number of neck and flank stripes, $\bar{X}$s males and females, respectively = 22.6, 23.1, t = −1.889, df = 198, p = 0.060; average stripe width on neck and flank, $\bar{X}$s = 56.9, 55.1, t = 0.730, df = 198, NS; ratio of widest black and white stripes, $\bar{X}$s = 0.61, 0.60, t = 1.351, df = 198, NS; percentage of neck black, $\bar{X}$s = 57.6%, 58.2%, t = −0.386, df = 198, NS; percentage of flank black, $\bar{X}$s = 47.7%, 48.5%, t = −0.509, df = 198, NS; percentage of rump black, $\bar{X}$s = 55.3%, 55.6%, t = −0.200, df = 84.141, NS.

[6] t-test comparing total amount of black pelage on rump of adult male and female zebras (N = 48, 152, respectively), $\bar{X}$s = 41.7, 43.8 pixels, t = −1.117, df = 198, NS.

[7] Pearson correlation coefficients between measures of striping and calculated shoulder heights and calculated weights, respectively (N = 288 individuals). Total number of neck and flank stripes, r = 0.004, NS, r = 0.046, NS; average stripe width on neck and flank, r = 0.014, NS, r = 0.113, p = 0.056; ratio of widest black and white stripes, r = 0.005, NS, r = 0.035, NS; percentage of neck black, r = −0.007, NS, r = 0.073, NS; percentage of flank black, r = −0.068, NS, r = −0.086, NS; percentage of rump black, r = 0.084, NS, r = 0.153, p = 0.009.

[8] t-tests comparing striping patterns of symmetrical and asymmetrical zebras when viewed from the front (N = 47, 92, respectively), total number of neck and flank stripes, t = −0.040, df = 111.431, NS; average stripe width on neck and flank, t = −0.195, df = 71.213, NS; ratio of widest black and white stripes, t = −0.750, df = 137, NS; percentage of neck black, t = −0.355, df = 137, NS; percentage of flank black, t = 0.185, df = 137, NS; percentage of rump black, t = −1.585, df = 72.441, NS.

[9] t-tests comparing striping patterns of symmetrical and asymmetrical zebras when viewed from the rear (N = 131, 82, respectively), total number of neck and flank stripes, t = 0.124, df = 211, NS; average stripe width on neck and flank, t = −0.606, df = 211, NS; ratio of widest black and white stripes, t = −1.014, df = 211, NS; percentage of neck black, t = −0.725, df = 211, NS; percentage of flank black, t = −0.416, df = 211, NS; percentage of rump black, t = −0.499, df = 211, NS.

[10] t-tests comparing striping patterns of zebras without wounds and those with wounds (N = 121, 73, respectively), total number of neck and flank stripes, t = −0.807, df = 192, NS; average stripe width on neck and flank, t = −1.904, df = 192, NS; ratio of widest black and white stripes, t = −1.219, df = 188.606, NS; percentage of neck black, t = −0.519, df = 192, NS; percentage of flank black, t = −0.358, df = 192, NS; percentage of rump black, t = 0.860, df = 192, NS.

Figure 7.0. A plains zebra standing in the open in the afternoon heat. Stripes have been conjectured to cool the animal by setting up convection currents over the body. (Drawing by Sheila Girling.)

SEVEN

Temperature regulation

7.1 Black and white surfaces

In studies of external coloration, it is important to distinguish between primary and secondary functions. Although feather, hair, or integument color may have been selected to attract mates or to signal dominance or to help in background matching, it will also have consequences for the animal's body temperature. Dark-colored pelage has greater absorptivity of shortwave radiation and acquires greater heat loads from solar radiation than lighter-colored pelage (Hamilton and Heppner 1967; Hamilton 1973). Thus, a priori, we would expect a zebra with large total expanses of white pelage to be cooler than an all-black alternative. Over and above this, however, there exists a cooling hypothesis for stripes that goes beyond white pelage secondarily cooling the animal. This states that the primary function of alternating black and white stripes is to set up convection currents around but particularly above the zebra, because black stripes absorb heat whereas white stripes reflect it, and that this process serves to cool the zebra (Cloudsley-Thompson 1984; Kingdon 1984; Morris 1990; Louw 1993). Observations using infrared photography do confirm that black stripes are warmer than white stripes and that the difference between them increases with rising air temperature. At night, however, temperature differences are reversed, with black stripes being cooler than white ones (Cena and Clark 1973; Benesch and Hilsberg 2003; Benesch and Hilsberg-Merz 2006).

This is a novel idea because it might also apply to other white and black mammals such as the indri or Angolan black and white colobus, although, as far as I am aware, it has never been advanced for any other black and white mammal (Caro 2009). Moreover, it is important to remember that absorption of solar radiation does not depend solely on coat color in mammals but also on structural characteristics of the coat, including hair density, hair lengths, diameters, angle, widths, and the air trapped beneath hairs—all of which can affect how much heat is transmitted to the skin surface (Cena and Monteith 1975; Burtt 1981; Maia, da Silva, and Bertipaglia 2005; Erdsack, Dehnhardt, and Hanke 2013). So a hypothesis that links hair color to temperature management needs to be tempered with information about hair structure too. Unfortunately, experiments that examine whether convection currents are generated by black and white striped pelage have not been performed, nor is it known if zebras are cooler than nonstriped equids. Here I compare the temperatures of zebras to those of nonstriped sympatric herbivores.

7.2 Heat measurements in the field

I used a ThermaCAM® EX320 FLIR infrared camera to capture images of four species of ungulates in Katavi National Park in February 2007 (wet season) and in July and August 2008 (dry season) in nonwindy conditions. I approached single animals or groups of individuals in a vehicle, stopped, pointed the camera at an individual zebra, impala, buffalo, or giraffe using X4 magnification. Thermal measurements were taken at some distance from individual animals (all species: N = 167 individuals, $\bar{X}$ = 51.6 m, min 13 m, max 196 m; zebras: N = 56, $\bar{X}$ = 68.9 m, min 25 m, max 142 m), and so for zebras readings constituted that reflected from the body, not individual black and white stripes. The camera was set at a standard 0.89 emissivity in the wet season and 0.98 in the dry season; emissivity is a measure of the efficiency with which a surface emits thermal energy compared to a black body that has an emissivity value of 1. Reflected temperature was set at a standard 20.0°C, atmospheric temperature at 20.0°C, relative humidity at 30.0%, and distance at 2.0 m. These initial measurements were not correct but were exactly the same across image captures. The camera generated a record of the minimum, maximum, and mean temperature of the whole image captured by the camera viewfinder, which I duly noted. Subsequently, I used the maximum temperature, as that was always a point on the animal, whereas the other measurements referred to the whole frame, which included the background (Plate 7.1).

Having taken the photograph, I recorded a series of variables that influence measurements of heat given off a surface. I used a thermometer to record the atmospheric temperature in the shade of the car. Using a different dry/wet bulb thermometer, I calculated relative humidity at the time. Finally, I recorded the distance from the vehicle to the animal using a range finder. Having noted these variables for each image, I then deleted it from the camera or retained it and subsequently downloaded it onto a laptop computer. There, using ThermaCAM® QuickView software, I reset object parameters to be 0.98 emissivity, retained the reflected temperature at an arbitrary 20.0°C, set the atmospheric temperature and relative humidity and distance to that recorded in the field for that photograph, and then read off the adjusted maximum temperature on the animal's body on the screen. In addition, I took the average temperature of a rectangle placed over the individual's flank. I also recorded the temperature at a single point on the head, neck, shoulder, middorsum, rump, stomach, and leg of the animal. For both the maximum temperatures read directly off the camera, and for all measurements read off the laptop, which were adjusted for correct object parameters, I used atmospheric temperature, relative humidity, distance, and whether the animal was in full sun (versus in the shade or standing under cloudy conditions) as covariates in analyses between species. It is worth noting that recorded temperatures do not represent true temperatures because reflected temperature was set at 20.0°C, which may be incorrect, and because the camera senses the relative temperature of hair rather than the skin surface or the animal's core temperature (Cena and Clark 1973). Nonetheless, infrared photography provides a good first approximation of external temperatures off different species taken under natural conditions (Plate 7.1).

I found no evidence that maximum temperatures of plains zebras were lower than those of impalas with their light brown coats, buffalo with their dark gray or black coats, or giraffes with mottled pelage (Figure 7.1). In

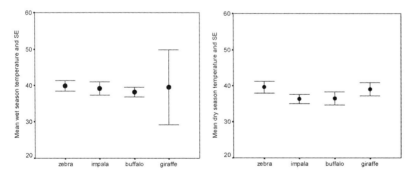

Figure 7.1. Mean and SE maximum temperatures (see text) (in degrees centigrade) in wet (left) and dry (right) seasons of four species in Katavi National Park. Ns = zebra 31, 25, impala 22, 27, buffalo 22, 13, giraffe 3, 24 respectively.

Table 7.1. Averages of adjusted external temperatures (in degrees centigrade) of four species of ungulate.

	Zebra	Impala	Buffalo	Giraffe
Body maxima	38.9	**36.6**	**36.4**	38.6
	(N = 57)	(N = 49)	(N = 35)	(N = 27)
Mean flank	36.0	**34.2**	**34.1**	**33.3**
	(N = 57)	(N = 49)	(N = 35)	(N = 27)
Head	34.7	**33.9**	**32.7**	**33.6**
	(N = 57)	(N = 49)	(N = 35)	(N = 27)
Neck	35.5	**33.2**	**33.2**	**33.3**
	(N = 57)	(N = 49)	(N = 35)	(N = 27)
Shoulder	37.1	**34.9**	**35.2**	**34.7**
	(N = 57)	(N = 49)	(N = 35)	(N = 27)
Back	37.1	**35.3**	**35.1**	35.1
	(N = 57)	(N = 49)	(N = 35)	(N = 24)
Rump	35.5	35.1	**34.0**	33.8
	(N = 57)	(N = 49)	(N = 35)	(N = 24)
Belly	33.6	**32.3**	**31.0**	**31.0**
	(N = 57)	(N = 49)	(N = 35)	(N = 23)
Leg	32.2	**31.1**	**29.9**	**30.5**
	(N = 56)	(N = 47)	(N = 34)	(N = 23)

Note: N denotes numbers of individuals; bold type shows recorded temperatures significantly ($P < 0.005$) lower than zebras.

the wet season, there was no significant difference between species' maximum recorded temperatures, although there was in the dry season [1]. In that season zebras' maximum temperatures were significantly greater than those of impalas and buffalo, but they did not differ significantly from those of giraffes [2].

Turning now to adjusted temperatures, again maximum temperatures were greater for zebras than impalas or buffalo, but not for giraffes (Table 7.1) [3]. Mean flank temperatures were significantly greater for zebra than all other species [4]. This is an important measure because it averages a considerable portion of the animals' flanks viewed laterally (Figure 7.2). Comparing specific areas of the body, there was not a single instance in which zebras were cooler than heterospecifics (Table 7.1). Rather, zebras were significantly warmer than impalas and buffalo and giraffes on virtually every part of their body except giraffes on their backs and rumps and impalas on their rumps [5–11]. It is noteworthy that zebras have significantly greater head temperatures than other species given the importance of keeping the brain cool (see below). Similarly, greater neck, dorsal, and rump temperatures are also notable given that the cooling hypothesis predicts exactly the opposite because it relies on convection currents being set up above the animal. My data categorically refute the idea that zebra

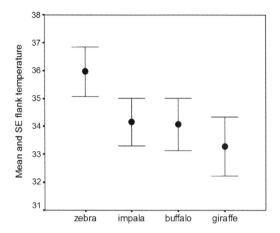

Figure 7.2. Mean and SE flank temperatures (in degrees centigrade) of four species in Katavi National Park. Ns = zebra 57, impala 49, buffalo 35, giraffe 27.

external temperatures are lower than those of sympatric nonstriped herbivores. One could certainly argue that zebras would be even warmer if they were entirely black, but it is difficult to argue from these data that striping per se serves to cool them.

In apparent contrast to these results, a multifactorial analysis conducted by Larison, Harrigan, Thomassen, and colleagues (2015) found a positive association between the degree of striping in 16 populations of plains zebras and temperature. As discussed in Chapter 8, they focused their attention on forelegs, hindlegs, torso, and belly and for each region scored stripe number, thickness, length, color saturation, and definition (i.e., standardized values of thickness × length × saturation) and matched these to a suite of environmental variables taken from WorldClim and other sources. They found several associations, the most prominent of which were isothermality, a measure of daily and annual temperature constancy, and mean temperatures during the coolest quarter of the year. While black and white stripes setting up convection currents is one explanation for this association, it would likely apply principally to the dorsum and perhaps the torso (Larison, Harrigan, Rubenstein, and Smith 2015) but not, as the authors acknowledge, to forelegs and hindlegs. Instead, the intraspecific association may be related to other variables, which include higher temperatures driving biting fly activity, or increasing disease infection rates of biting flies, or of zebras, or interactions among these variables.

Unpublished data by Irondo and Rubenstein mentioned in the Larison, Harrigan, Thomassen, and colleagues (2015) paper report that zebras maintain significantly lower surface body temperatures (29.2°C) than similar-size herbivores (32.5°C) (these were principally hartebeest [D. Rubenstein, personal communication]). Irondo and Rubenstein used a laser infrared digital thermometer gun, whereas I used an infrared camera, and our disparate results need to be clarified. Whatever the outcome, it is worth remembering that reflected surface temperatures may not represent internal temperatures and that large expanses of white pelage, a priori, are likely to result in less environmental heat gain at the skin and less evaporative water loss compared to dark pelage (Finch, Bennett, and Holmes 1984), so that the driver of any cooling may not be convection currents but white reflective hair. Moreover, convective cooling can also occur without striping: dark steers show high convective cooling resulting from high coat surface temperatures (Hutchinson and Brown 1969), although it is noteworthy that this does not counterbalance their heat load due to absorption of solar radiation (Finch et al. 1980; Finch, Bennett, and Holmes 1984).

In passing, it is worth noting that zebras' skins are thicker than those of other unstriped equids, at least as measured from museum pelts [12], and are also thicker than a suite of sympatric artiodactyl species [13] (Table 7.2). It is difficult to know how this might influence thermoregulation. On the one hand, one could argue that because zebras have a thick epidermis, they need additional mechanisms (such as possible convection currents produced by stripes) to stay cool, although my infrared thermocamera data contradict this. On the other hand, one could argue that a thicker epi-

Table 7.2. Skin thicknesses of herbivores in cm measured from museum pelts.

	N	$\bar{X}$	SD
Plains zebra	15	1.42	0.29
Mountain zebra	6	1.61	0.15
Grevy's zebra	8	1.63	0.17
Asiatic wild ass	7	1.16	0.09
Przewalski's horse	2	1.69	0.62
Blue wildebeest	8	1.34	0.25
Roan antelope	10	1.38	0.19
Oryx	9	1.47	0.36
Hartebeest	7	1.20	0.27
Waterbuck	9	1.40	0.18
Topi	13	1.35	0.15

Source: H. Walker, unpublished data.
Notes: Data represent average of neck, middorsum, rump, center of buttock, and center of belly measurements, and then averaged across individuals.

dermal layer means that zebras do not suffer heat stress and that there is no need to employ cooling morphology. Zebras wear shallow coats (Chapter 8) that provide low thermal insulation. Such coats dissipate heat well from those parts of the body not exposed to radiation and at times of the day when a zebra is not exposed to radiation (Hutchinson and Brown 1969). The whole issue of zebra skin and hair morphology and their interactions with hair coloration needs thorough investigation, and comparisons between striped and unstriped dyed zebras would be useful.

7.3 Heat management

Animals must not overheat whatever their pelage coloration, so many tropical species move to shade during the heat of the day or else cease movement if shade is unavailable. Behavioral regulation of body temperature might be especially important for free-living zebras because they are incapable of selective brain cooling. Brain temperatures may exceed blood temperatures by 0.2–0.4°C, reaching 40.5°C (Fuller et al. 2000), jeopardizing cerebral function and exacerbating water loss in arid environments (Kuhnen 1997; Jessen et al. 1998). As plains and mountain zebras have broad bands of white pelage that must reflect heat and that can be as much as 9°C cooler than corresponding areas of black pelage (Cena and Clark 1973), and as white pelage is found particularly on the rump, inside of the legs, and on the bellies of some subspecies, I expected that zebras might use the white areas of their bodies to manage temperature (as noted in other African herbivores; Hetem et al. 2011). Specifically, I expected that zebras might present their rumps to the sun to gain maximum reflectance during hotter periods of the day under nonwindy conditions; radiative heat load transferred to the skin is less for white than black surfaces under such circumstances (Walsberg, Campbell, and King 1978).

To test this idea, I noted the direction in which zebras were standing during those sessions in which I was trying to ascertain the extent of mutual grooming (section 6.3.a). I discovered that individuals were more likely to direct their rumps toward the sun (an average of 44.5% of 209 individuals during 26 three-hour sessions) than their faces (10.5%, Ns = 209 individuals, 26 sessions) or other parts of their body toward it (42.8%, Ns = 204 individuals, 25 sessions). I also recorded zebras' body orientations in relation to the sun across seven afternoon observation periods between 16.20 hours and 18.00 hours. Every 10 minutes I recorded the proportion of individuals in a group that presented their rumps to the sun as well as the temperature, taken from the shade of the vehicle. On most afternoons,

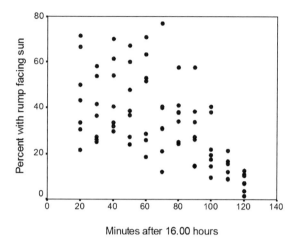

Minutes after 16.00 hours

Figure 7.3. Proportion of individual zebras in groups that directed their rumps to the sun plotted against minutes after 16.00 hours (N = 77 zebra—10-minute interval periods).

I found that the proportions of individuals that pointed their rumps to the sun were significantly positively correlated with temperature [14]. During the midafternoon heat, many individuals would stand motionless turned away from the bright sun, and there was a clear change in body orientation from midafternoon to evening [15] (Figure 7.3).

7.4 Conclusions

The thermographic measurements presented in this chapter show no evidence of zebras being cooler than sympatric ungulates. At first sight this is somewhat surprising as nearly half of the external pelage of zebras consists of white hair that reflects shortwave radiation well, but this may simply be countered by the other half being black. My findings that zebras are no cooler than sympatric ungulates are the strongest evidence against the primary function of stripes being for cooling, but they are just one of four separate lines of argument. First, design features of stripes speak against them having evolved to keep individuals cool: black stripes cover a greater area of the body than white stripes, at least in plains zebras (Table 6.4), and hence present a considerable area for heat absorption. If zebra pelage coloration were selected to cool the animal, we might expect more of a preponderance of white pelage. If black and white stripes were a cooling mechanism, why are these replaced by black skin around the nostrils and mouth? Given that air taken in through these orifices will be warmer as a consequence of this area being black, it is surprising under the cooling hypothesis that white and black hairs do not extend farther forward. An additional two points are that in many subspecies of zebras, black lateral stripes are thicker dorsally and thinner ventrally, and that pelage directly along a zebra's spine often consists of a black stripe that connects up with lateral black stripes along the flanks. If striping served to cool the zebra, it is surprising that this part of the dorsum is so dark, since the overhead midday sun falls on it. In the mountain zebra, the black stripes form a thick gridiron on the animal's rump, which again must absorb heat during the hottest part of the day.

Second, there are environmental considerations: zebras out on the plains often encounter windy conditions in open habitat, making it difficult to see how convection currents could help cool an animal substantially under these conditions (see Walsberg 1983). Furthermore, eddies over the animal's surface will dissipate as soon as the animal moves (D. Rubenstein, personal communication). As an anecdote, I have never seen dust eddies form across a zebra's dorsum despite many hours of watching zebras in dusty conditions.

Third, there are comparative arguments against stripes functioning as heat managers. Many taxa that inhabit very hot treeless environments are pale gray, pale yellow, or cream colored, presumably to reflect sunlight (Stoner, Caro, and Graham 2003; but see Hutchinson and Brown 1969; Oristland 1970). For instance, thermal measurements of impala using implanted thermometers show that light-colored coats reduce overheating during hot times of the year (Hetem et al. 2009). Light-colored desert species include the addax, scimitar-horned oryx, carnivores like the fennec, and open-country equids such as African and Asiatic wild asses. Logically, then, the primary function of stripes is unlikely to be thermal in nature, as closely related species apparently sport light coats to remain cool. Another comparative argument runs as follows. Zebras are open-country grazers often bereft of shade, but so are sympatric wildebeest, eland, and topi, to name just a few species. None of these have particularly light coats, and none are striped. Absence of light pelage in these species suggests that the open plains of eastern and southern Africa are insufficiently hot to warrant special cooling mechanisms derived from coat color; de facto striping constitutes no such mechanism. Yet another comparative argument concerns equids only. Grevy's and plains zebras graze in in open areas, so they may be subject to heat stress in the middle of the day, especially during the dry season. Yet the mountain zebra lives in cooler areas of South Africa but is nonetheless striped. Furthermore, the unstriped African wild ass and subspecies of the Asiatic wild ass in Iran, India, and formerly in Syria live in very hot regions of the Old World, but none of these species are striped (Morris 1990). I explore this issue systematically in Chapter 8. These three lines of argument together with the data presented here refute stripes being an evolutionary response to heat stress, but there must be thermal secondary consequences of this extraordinary pelage, so it is likely that there is more to discover regarding temperature regulation and striping.

For example, white stripes may be used to reduce heat load. My observations on body orientation being related to temperature during the heat of the day suggest this but may conceivably be interpreted in other ways, including shielding the eyes from direct bright light (Pennacchio, Cuthill, et al. 2015). Discriminating among such possibilities requires comparisons with body orientation of unstriped ungulates in bright afternoon light, and although I have no data on this, my informal observations indicate that sympatric topi do not stand with their hindquarters to the sun. This suggests that white rumps may be employed to ameliorate heat load. Conversely, blacker areas of the body—the head and neck, for instance—may be presented to the sun in order to warm up in the mornings, although I have no data on this.

Statistical tests

[1] Wet season comparisons of maximum temperatures of zebra (N = 31, $\bar{X}$ = 39.9°C), impala (N = 22, $\bar{X}$ = 39.2°C), buffalo (N = 22, $\bar{X}$ = 38.2°C), and giraffe (N = 3, $\bar{X}$ = 39.5°C). Model F = 4.980, df = 7, 70, p < 0.0001; atmospheric temperature F = 1.874, df = 1, 70, NS; relative humidity F = 0.264, df = 1, 70, NS; distance F = 2.312, df = 1, 70, NS; sun F = 14.056, df = 1, 70, p < 0.0001; species F = 1.186, df = 3, 70, NS.

Dry season comparisons of maximum temperatures of zebra (N = 25, $\bar{X}$ = 39.9°C), impala (N = 27, $\bar{X}$ = 36.3°C), buffalo (N = 13, $\bar{X}$ = 36.5°C), and giraffe (N = 24, $\bar{X}$ = 39.0°C). Model F = 7.762, df = 7, 81, p < 0.0001; atmospheric temperature F = 16.951, df = 1, 81, p < 0.0001; relative humidity F = 2.656, df = 1, 81, NS; distance F = 3.458, df = 1, 81, p = 0.067; sun F = 0.774, df = 1, 81, NS; species F = 5.831, df = 3, 81, p = 0.001.

[2] Dry season maximum temperatures: Zebra vs. impala: model F = 8.789, df = 5, 46, p < 0.0001; atmospheric temperature F = 8.431, df = 1, 46, p = 0.006; relative humidity F = 1.727, df = 1, 46, NS; distance F = 4.435, df = 1, 46, p = 0.041; sun F = 0.650, df = 1, 46, NS; species F = 15.316, df = 1, 46, p < 0.0001. Zebra vs. buffalo: model F = 4.175, df = 5, 32, p = 0.005; atmospheric temperature F = 2.738, df = 1, 32, NS; relative humidity F = 0.040, df = 1, 32, NS; distance F = 1.499, df = 1, 32, NS; sun F = 0.316, df = 1, 32, NS; species F = 9.121, df = 1, 32, p = 0.005. Zebra vs. giraffe: model F = 5.845, df = 5, 43, p < 0.0001; atmospheric temperature F = 6.952, df = 1, 43, p = 0.012; relative humidity F = 0.194, df = 1, 43, NS; distance F

= 0.089, df = 1, 43, NS; sun F = 6.420, df = 1, 43, p = 0.015; species F = 0.801, df = 1, 43, NS.

[3] Wet and dry season combined adjusted maximum temperatures: Zebra vs. impala: model F = 8.502, df = 5, 99, p < 0.0001; atmospheric temperature F = 12.164, df = 1, 99, p = 0.001; relative humidity F = 2.251, df = 1, 99, NS; distance F = 4.435, df = 1, 99, p = 0.064; sun F = 6.856, df = 1, 99, p = 0.010; species F = 8.145, df = 1, 99, p = 0.005. Zebra vs. buffalo: model F = 6.668, df = 5, 85, p < 0.0001; atmospheric temperature F = 4.093, df = 1, 85, p = 0.046; relative humidity F = 0.101, df = 1, 85, NS; distance, F = 1.660, df = 1, 85, NS; sun F = 5.656, df = 1, 85, p = 0.020; species F = 9.561, df = 1, 85, p = 0.003. Zebra vs. giraffe: model F = 9.663, df = 5, 77, p < 0.0001; atmospheric temperature F = 8.850, df = 1, 77, p = 0.004; relative humidity F = 0.012, df = 1, 77, NS; distance F = 0.002, df = 1, 77, NS; sun F = 20.727, df = 1, 77, p < 0.0001; species F = 0.010, df = 1, 77, NS.

[4] Wet and dry season combined adjusted mean flank temperatures: Zebra vs. impala: model F = 9.270, df = 5, 99, p < 0.0001; atmospheric temperature F = 11.052, df = 1, 99, p = 0.001; relative humidity F = 17.919, df = 1, 99, p < 0.0001; distance F = 6.980, df = 1, 99, p = 0.010; sun F = 10.823, df = 1, 99, p = 0.001; species F = 10.420, df = 1, 99, p = 0.002. Zebra vs. buffalo: model F = 4.563, df = 5, 85, p = 0.001; atmospheric temperature F = 3.165, df = 1, 85, p = 0.079; relative humidity F = 2.283, df = 1, 85, NS; distance F = 1.939, df = 1, 85, NS; sun F = 5.624, df = 1, 85, p = 0.020; species F = 9.242, df = 1, 85, p = 0.003. Zebra vs. giraffe: model F = 10.228, df = 5, 77, p < 0.0001; atmospheric temperature F = 6.108, df = 1, 77, p = 0.016; relative humidity F = 3.618, df = 1, 77, p = 0.061; distance F = 0.937, df = 1, 77, NS; sun F = 19.433, df = 1, 77, p < 0.0001; species F = 11.888, df = 1, 77, p = 0.001.

[5] Wet and dry season combined adjusted head temperatures: Zebra vs. impala: model F = 10.855, df = 5, 99, p < 0.0001; atmospheric temperature F = 13.658, df = 1, 99, p < 0.0001; relative humidity F = 13.869, df = 1, 99, p < 0.0001; distance F = 20.554, p < 0.0001; sun F = 8.296, df = 1, 99, p = 0.005; species F = 12.091, df = 1, 99, p = 0.001. Zebra vs. buffalo: model F = 7.386, df = 5, 85, p < 0.0001; atmospheric temperature F = 3.379, df = 1, 85, p = 0.070; relative humidity F = 0.560, df = 1, 85, NS; distance F = 6.863, df = 1, 85, p = 0.010; sun F = 3.056, df = 1, 85, p = 0.084; species F = 17.648, df = 1, 85, p < 0.0001. Zebra vs. giraffe: model F = 9.301, df = 5, 77, p < 0.0001; atmospheric temperature F = 16.961, df = 1, 77, p < 0.0001; relative humidity F = 3.385, df = 1, 77, p = 0.070; distance F = 4.768, df = 1, 77, p = 0.032; sun F = 9.230, df = 1, 77, p = 0.003; species F = 5.393, df = 1, 77, p = 0.023.

[6] Wet and dry season combined adjusted neck temperatures: Zebra vs. impala: model F = 8.014, df = 5, 99, p < 0.0001; atmospheric temperature F = 5.474, df = 1, 99, p = 0.021; relative humidity F = 7.928, df = 1, 99, p = 0.006; distance F = 5.480, df = 1, 99, p = 0.021; sun F = 6.831, df = 1, 99, p = 0.010; species F = 15.927, df = 1, 99, p < 0.0001. Zebra vs. buffalo: model F = 4.554, df = 5, 85, p = 0.001; atmospheric temperature F = 0.435, df = 1, 85, NS; relative humidity F = 0.039, df = 1, 85, NS; distance F = 0.626, df = 1, 85, NS; sun F = 3.432, df = 1,

85, p = 0.067; species F = 13.216, df = 1, 85, p < 0.0001. Zebra vs. giraffe: model F = 6.285, df = 5, 77, p < 0.0001; atmospheric temperature F = 1.711, df = 1, 77, NS; relative humidity F = 0.256, df = 1, 77, NS; distance F = 0.304, df = 1, 77, NS; sun F = 10.421, df = 1, 77, p = 0.002; species F = 9.350, df = 1, 77, p = 0.003.

[7] Wet and dry season combined adjusted shoulder temperatures: Zebra vs. impala: model F = 7.640, df = 5, 99, p < 0.0001; atmospheric temperature F = 10.186, df = 1, 99, p = 0.002; relative humidity F = 11.978, df = 1, 99, p = 0.001; distance F = 2.559, df = 1, 99, NS; sun F = 9.709, df = 1, 99, p = 0.002; species F = 6.522, df = 1, 99, p = 0.012. Zebra vs. buffalo: model F = 3.319, df = 5, 85, p = 0.009; atmospheric temperature F = 1.536, df = 1, 85, NS; relative humidity F = 0.519, df = 1, 85, NS; distance F = 0.872, df = 1, 85, NS; sun F = 4.927, df = 1, 85, p = 0.029; species F = 5.883, df = 1, 85, p = 0.017. Zebra vs. giraffe: model F = 7.250, df = 5, 77, p < 0.0001; atmospheric temperature F = 3.589, df = 1, 77, p = 0.062; relative humidity F = 0.654, df = 1, 77, NS; distance F = 0.104, df = 1, 77, NS; sun F = 16.287, df = 1, 77, p < 0.0001; species F = 5.450, df = 1, 77, p = 0.022.

[8] Wet and dry season combined adjusted dorsum temperatures: Zebra vs. impala: model F = 5.737, df = 5, 99, p < 0.0001; atmospheric temperature F = 7.967, df = 1, 99, p = 0.006; relative humidity F = 6.040, df = 1, 99, p = 0.016; distance F = 3.007, df = 1, 99, p = 0.086; sun F = 7.086, df = 1, 99, p = 0.009; species F = 5.249, df = 1, 99, p = 0.024. Zebra vs. buffalo: model F = 3.544, df = 5, 85, p = 0.006; atmospheric temperature F = 2.799, df = 1, 85, p = 0.098; relative humidity F = 0.403, df = 1, 85, NS; distance F = 0.553, df = 1, 85, NS; sun F = 4.678, df = 1, 85, p = 0.033; species F = 5.812, df = 1, 85, p = 0.018. Zebra vs. giraffe: model F = 6.529, df = 5, 74, p < 0.0001; atmospheric temperature F = 4.448, df = 1, 74, p = 0.038; relative humidity F = 0.010, df = 1, 74, NS; distance F = 0.120, df = 1, 74, NS; sun F = 13.477, df = 1, 74, p < 0.0001; species F = 1.637, df = 1, 74, NS.

[9] Wet and dry season combined adjusted rump temperatures: Zebra vs. impala: model F = 7.189, df = 5, 99, p < 0.0001; atmospheric temperature F = 11.675, df = 1, 99, p = 0.001; relative humidity F = 14.686, df = 1, 99, p <0.0001; distance F = 8.856, df = 1, 99, p = 0.004; sun F = 9.926, df = 1, 99, p = 0.002; species F = 2.462, df = 1, 99, NS. Zebra vs. buffalo: model F = 3.030, df = 5, 85, p = 0.015; atmospheric temperature F = 3.489, df = 1, 85, p = 0.065; relative humidity F = 1.269, df = 1, 85, NS; distance F = 0.433, df = 1, 85, NS; sun F = 3.713, df = 1, 85, p = 0.057; species F = 5.421, df = 1, 85, p = 0.022. Zebra vs. giraffe: model F = 6.380, df = 5, 74, p < 0.0001; atmospheric temperature F = 3.317, df = 1, 74, p = 0.073; relative humidity F = 0.494, df = 1, 74, NS; distance F = 0.433, df = 1, 74, NS; sun F = 13.753, df = 1, 74, p < 0.0001; species F = 3.934, df = 1, 74, p = 0.051.

[10] Wet and dry season combined adjusted belly temperatures: Zebra vs. impala: model F = 8.226, df = 5, 99, p < 0.0001; atmospheric temperature F = 11.357, df = 1, 99, p = 0.001; relative humidity F = 1.349, df = 1, 99, NS; distance F = 3.161, df = 1, 99, p = 0.078; sun F = 8.024, df = 1, 99, p = 0.006; species F = 5.844, df = 1, 99, p = 0.017. Zebra vs. buffalo: model F = 16.614, df = 5, 85, p < 0.0001; atmospheric temperature F = 1.329, df = 1, 85, NS; relative humidity F = 4.241, df = 1, 85, p = 0.043; distance F = 2.431, df = 1, 85, NS; sun F = 8.738, df = 1, 85,

p = 0.004; species F = 32.983, df = 1, 85, p < 0.0001. Zebra vs. giraffe: model F = 13.896, df = 5, 73, p < 0.0001; atmospheric temperature F = 7.232, df = 1, 73, p = 0.009; relative humidity F = 1.496, df = 1, 73, NS; distance F = 0.006, df = 1, 73, NS; sun F = 8.609, df = 1, 73, p = 0.004; species F = 23.207, df = 1, 73, p < 0.0001.

[11] Wet and dry season combined adjusted leg temperatures: Zebra vs. impala: model F = 16.537, df = 5, 96, p < 0.0001; atmospheric temperature F = 14.946, df = 1, 96, p < 0.0001; relative humidity F = 0.059, df = 1, 96, NS; distance F = 8.940, df = 1, 96, p = 0.004; sun F = 14.734, df = 1, 96, p < 0.0001; species F = 7.751, df = 1, 96, p = 0.006. Zebra vs. buffalo: model F = 19.954, df = 5, 83, p < 0.0001; atmospheric temperature F = 7.814, df = 1, 83, p = 0.006; relative humidity F = 7.604, df = 1, 83, p = 0.007; distance F = 2.626, df = 1, 83, NS; sun F = 11.823, df = 1, 83, p = 0.001; species F = 19.709, df = 1, 83, p < 0.0001. Zebra vs. giraffe: model F = 15.378, df = 5, 72, p < 0.0001; atmospheric temperature F = 12.919, df = 1, 72, p = 0.001; relative humidity F = 1.843, df = 1, 72, NS; distance F = 2.498, df = 1, 72, NS; sun F = 7.518, df = 1, 72, p = 0.008; species F = 14.323, df = 1, 72, p < 0.0001.

[12] t-tests comparing overall skin thicknesses of individuals of 3 striped species of equids (N = 29 individuals, $\bar{X}$ = 1.52 cm) and individuals of 2 unstriped species of equids (N = 9, $\bar{X}$ = 1.28 cm), t = 2.263, df = 36, p = 0.030.

[13] t-tests comparing overall skin thicknesses of individuals of 3 striped species of equids (N = 29 individuals, $\bar{X}$ = 1.52 cm) and individuals of 6 unstriped sympatric artiodactyl species (N = 56, $\bar{X}$ = 1.36 cm), t = 2.736, df = 83, p = 0.008.

[14] Proportion of individual zebras that pointed their rumps to the sun measured every 10 minutes during late afternoons correlated with temperature (N = 11 scans each day). Day 1, r_s = + 0.708, p = 0.015; day 2, r_s = 0.521, p = 0.101; day 3, r_s = 0.815, p = 0.002; day 4, r_s = 0.631, p = 0.037; day 5, r_s = 0.588, p = 0.057; day 6, r_s = 0.977, p < 0.0001; day 7, r_s = 0.874, p < 00001.

[15] Proportion of individual zebras in groups that directed their rumps to the sun after 16.00 hrs correlated with time of the day, N = 77 zebra-10 minute interval periods, r_s = −0.632, p < 0.0001.

Figure 8.0. An Asiatic wild ass, or kulan (*above*), and a mountain zebra (*below*), both feeding. Comparisons between unstriped and striped equids are key to solving the riddle of why zebras have stripes. (Drawing by Sheila Girling.)

EIGHT

Multifactorial analyses

8.1 Comparing hypotheses simultaneously

The history of ideas about zebra coloration has been one of proposing a hypothesis and then discussing it or occasionally testing it. The possibility of testing several hypotheses simultaneously is new but can now be accomplished with the advent of geographic information technology. Instead of using observations or simple experiments to test notions about the functions of zebra stripes and examining one at a time, we can place several hypotheses in a multifactorial analysis that statistically titrates each one against each other. This is particularly important where explanatory factors are so geographically congruent—for example, tsetse flies, spotted hyenas, and woodlands cover roughly the same parts of the African continent where zebras live. In this chapter I discuss a comparative interspecific multifactorial analysis in equids and an intraspecific multifactorial analysis of striping in plains zebras, the only two analyses of this kind.

8.2 The interspecific comparison

A group of colleagues and I compared the seven extant wild equid species, some of which are striped and some of which are not (section 1.3), and further compared equid subspecies that show considerable variation in striping patterns too (Kingdon 1979; Groves and Bell 2004). We capitalized on this pelage variation by matching species' and subspecies' geographic range overlaps to several of the factors proposed for driving the evolution of zebras' coat coloration and did this simultaneously. The hypotheses we chose were (i) crypsis against a woodland background, (ii) thwarting predatory attack, (iii) avoiding ectoparasite attack, (iv) stripes having a social function, and (v) stripes reducing thermal load (Morris 1990; Cloudsley-Thompson 1999; Ruxton 2002). While possibly conspecific with *E. ferus przewalskii* (Ishida et al. 1995), we excluded domestic and feral horses because their coat color is labile and has been subject to intense selection through domestication (Ludwig et al. 2009). Amanda Izzo scored striping and collated range maps; Bobby Reiner digitized the range maps and quantified overlaps among them; Hannah Walker collected museum pelt data; and Ted Stankowich conducted statistical analyses and provided theoretical input during data collection and manuscript preparation (Caro et al. 2014).

Controlling for phylogeny both at the subspecies level as well as at the species level, we tested (i) whether black and white stripes are a form of crypsis, potentially matching the background of light and dark produced by tree trunks and branches in woodland habitats (Cott 1940). To investigate an antipredator function that subsumes several mechanisms discussed in Chapters 2, 3, and 4, which are crypsis, aposematism, and various aspects of confusion or pursuit deterrence (e.g., Galton 1851; Cott 1966; Ljetoff et al. 2007; Stevens, Yule, and Ruxton 2008), we simply examined (ii) whether striping in equids is associated with the presence of sympatric large predators: spotted hyena, lion, tiger, and wolf known to prey on equids. (iii) To investigate the importance of biting flies (e.g., R. Harris 1930; Lehane 2005; Egri, Blaho, Kriska, et al. 2012), we matched tsetse fly distributions to equid distributions and used environmental proxies for the distribution of tabanid abundance. (iv) We used average group size and maximum herd size as coarse indicators of the extent of social interactions between individuals (Kingdon 1979; Kingdon 1984; Morris 1990). (v) To explore heat management (Morris 1990; Louw 1993), we examined associations between striping and living in habitats with average maximum annual temperatures of 25°C through 30°C.

8.2.a Comparative methodology

Stripes and pelage background coloration

To measure equid pelage coloration, we analyzed coat patterns and colors quantitatively using images of extant and extinct equid subspecies obtained from reputable sources online and in print (see Appendix 5). Only one animal per photograph was scored, and we scored a maximum of 12 animals per subspecies (see Appendix 5 for Ns). Images were scored only if the animal was identified with the proper scientific and common name, and if the location where the image was taken was identified to occur within the species' (or subspecies') known range. In total, we scored the following species and subspecies (Moehlman 2002c): (i) Plains zebra (with six subspecies: Damara, *E. b. antiquorum*; Burchell's, *E. b. burchellii* [extinct]; Chapman's, *E. b. chapmani*; Crawshay's, *E. b. crawshayi*; Upper Zambezi, *E. b. zambeziensis*; Grant's, *E. b. boehmi*; and the extinct quagga, *E. b. quaggai*. (ii) Mountain zebra (with two subspecies: Hartmann's, *E. z. hartmannae*; and Cape, *E. z. zebra*). (iii) Grevy's zebra. (iv) African wild ass (with 3 subspecies: Somali, *E. a. somaliensis*; Nubian, *E. a. africanus*, and Atlas [extinct], *E. a. atlanticus*). (v) Asiatic wild ass (with six subspecies: Mongolian, *E. h. hemionus*; Gobi, *E. h. leuteus*; Kulan, *E. h. kulan*; Onager, *E. h. onager*; Indian, *E. h. khur*; and Syrian, *E. h. hemippus* [extinct]). (vi) Kiang (with three subspecies: Eastern, *E. k. holdereri*; Western, *E. k. kiang*; and Southern, *E. k. polyodon*). (vii) Przewalski's horse. Mean scores of photographs were calculated for each subspecies, and these scores were averaged again to generate species scores.

To quantify pelage coloration, we followed the methodology of Groves and Bell (2004) and used three of their criteria for each photograph. Briefly, (a) belly striping—namely, the number of stripes connected to the ventral midline of the animal; (b) shadow stripes: 0 = none, 1 = poorly developed, 2 = clear on rump (i.e., the area behind the flank) not on neck, 3 = strong on rump and neck; and (c) degree of leg stripes: 0 = none, 1 = none/traces below elbow and stifle, 2 = clearly present to knee/hock, 3 = down to hoof, but broken or incomplete, 4 = complete, unbroken. In addition, we developed several other categories to capture further the variation observed among various zebra subspecies. These were (d) the overall appearance of the color of the legs (scored as light, or medium, or dark), from which we created a new measure called leg stripe intensity—namely, the score for leg striping (0 through 4) multiplied by the color of the legs (1 for white/light, 2 for "black"/dark). This measure captures both the degree of striping and the darkness of leg coloration, and ranges from 0 to 8.0. (e) Facial stripes— namely, the number of black stripes on the animal's face, from the end of the solid-colored portion of the muzzle up to the throat. (f) Neck stripes, the number of black stripes from the animal's throat to the animal's withers. (g) Flank stripes—namely, the number of vertical body stripes found on the lateral surface (the area between the vertical lines from groin and withers to the elbow) of the animal. For the same reason, this was subsequently recoded into 4 broad categories to avoid minor differences in stripe number having undue influence: no stripes = 0, 1–2.9 stripes = 1, 3–6 stripes = 2, and >6 stripes = 3 and is called flank striping. (h) The number of horizontal stripes (counted) on the rump. This was subsequently recoded into four categories: no stripes = 0, 1–4.9 stripes = 1, 5–10 stripes = 2, and >10 stripes = 3 and is called rump striping. We elected to analyze striping patterns on different parts of the body because they varied considerably in width, number, and orientation (Penzhorn 2013), and we purposely used a variety of methods, including presence, number, and intensity of stripes, as we did not know to what predators, flies, or conspecifics attend.

Range maps

For the historical ranges of equid species and subspecies, we consulted Moehlman (2002a), except for Przewalski's horse, taken from Boyd (1988); for subspecies of the African wild ass, taken from Kimura and colleagues (2010); and for the kiang and Asiatic wild ass, taken from Groves (1974).

To measure degree of overlap among equid ranges and our independent variables—predator ranges, temperature ranges, vegetation zones, tsetse fly distributions, and proxies for tabanid fly distributions—we used the software ArcGIS 10. We started with a map of the Old World as the basemap (from Natural Earth, free vector and raster map data: naturalearthdata .com) onto which we overlaid the data. Data consisted of range maps for each species or subspecies, and our independent variables that we georeferenced onto the basemap using a minimum of four reference points, and distance errors were minimized to ensure accurate map overlap. Once the range maps were rasterized onto the basemap, polygons were created by ArcGIS; these polygons contain spatial information, such that the size of a particular range can be calculated as well as the amount of overlap among various polygons. We used ArcGIS 10 to determine areas of overlap between each combination of ranges. These areas of overlap were then used to create the percentage overlap of each subspecies' range with a given independent variable (using R), and the percentage overlaps are used as the data points.

To derive accurate overlap information for species distributions, we were careful to select reputable sources for our range information. Information on tsetse fly distribution came from the Food and Agriculture Organisation of the United Nations' *Consultants' Report: Predicted Distributions of Tsetse in Africa*, published in February 2000 (http://www.fao.org /ag/againfo/programmes/en/paat/documents/maps/pdf/tserep.pdf). We used historical presence/absence maps for the superfamily groups, *Glos-*

sina palpalis, *morsitans*, and *fusca*, which are the standard groups used for tsetse flies. ArcGIS 10 was then able to find the union of these three distributions to determine an overall tsetse distribution in Africa.

For presence or absence of predator species, we used historical range information because geographic ranges of large carnivores have contracted drastically in recent times and historical ranges are most likely to have influenced the evolution of equid coloration. We included the following predators: spotted hyena, whose historic range was taken from the IUCN (iucnredlist.org); lion, whose historic range was taken from Panthera (panthera.org); tiger, whose historic range was taken from Panthera (panthera.org); and wolf, whose historic range for Asia was taken from the IUCN (iucnredlist.org).

For world vegetation zones, we used a world biome map obtained from http://www.kidsmaps.com/geography/The+World/Economic/Biomes+of +the+World. We chose this source because (i) the website is run by Wikimedia, which only contains reviewed sources; (ii) map quality and resolution are high and lend themselves favorably to use in ArcGIS; and (iii) the map contains numerous detailed and specific vegetation categories, desirable for our goal of determining whether habitat impacts the evolution of zebra coloration.

Tabanidae are a diverse group comprising over 4154 species and are found globally except on oceanic islands and at the poles, but unfortunately only very general distribution maps exist (Mackerras 1954). In temperate zones they are abundant in summer, and in the tropics they principally appear during the wet season. More specifically, optimal tabanid habitat is thought to exist in locations where temperatures lie between 15° and 30°C and humidity is between 30% and 85% (Tashiro and Schwardt 1949; Miller 1951; Chvala, Lyneborg, and Moucha 1972; Burnett and Hays 1974; Alverson and Noblet 1977; Nilsson 1997). Because climate changes from month to month, locations that are in the optimal temperature and humidity ranges were identified for each month of the year. We utilized GIS products of maximum daily temperature and daily relative humidity at 3:00 p.m., aggregated over each month, generated as output by a recent global climatology effort that aggregates and adjusts historical climate data from across the world (Kriticos et al. 2011). Specifically, we used data on monthly values of humidity and maximum daily temperature that had a resolution of 10 feet squared. We combined the two data files to identify, for each month, which locations were in the optimal range.

There is significant intra-annual variation in climate, and as such, regions that are optimal for tabanid activity in July are not necessarily optimal in January. Many locations are optimal for a single month, but few are optimal all year round. We identified which locations were frequently optimal and which were rarely optimal. Specifically, we created 12 distinct maps, each one identifying which locations are optimal at least that many months in a row (e.g., the fourth map identifies locations that have 4 consecutive months in the optimal range). The first map shows a large range where only one month in the optimal range is needed, while the twelfth map shows the very small range where the temperature and humidity are optimal all year long. All analyses were performed in ArcGIS 10 (to read the initial map files and identify the individual optimal regions) and R 2.15.0 using programs such as maptools, rgeos, and sp (to combine the temperature and humidity maps and construct the 12 distinct maps).

All distribution maps of temperature isoclines were obtained from the National Oceanic and Atmospheric Administration (NOAA; www.noaa .gov).

Group sizes

Mean adult group sizes and maximum herd sizes for each species (but not for subspecies) were taken from the literature, and for these comparative analyses we used the following figures: plains zebra, mean 5.6 (Grubb 1981), maximum 16 (Linklater 2000); mountain zebra, 4.7 (Penzhorn and Novellie 1991), 13 (Linklater 2000); Grevy's zebra, 5.1 (Sundaresan, Fischhoff, Dushoff, et al. 2007), 150 (S. Williams 2013); African wild ass, 4.0, 49 (Klingel 1977); Asiatic wild ass, 6.3, 850 (Feh et al. 2001); Kiang, 8.9, 500 (St-Louis and Cote 2009); Przewalski's horse, 7.0, 20 (Linklater 2000).

For derivation of phylogenies and phylogenetic analyses, see Appendixes 6 and 7, respectively.

8.2.b Overall striping

Starting with just single factor analyses at the species level [1–7], presence of body stripes was very strongly associated with presence of tsetse flies, with several measures of tabanid fly activity, and with presence of lions. For geographic distributions of tsetse flies, and for five, six, seven, eight, nine, ten, and eleven consecutive months of good conditions for tabanids, and for presence of lions, the striped equid species with the lowest geographic range overlap on each of these measures still had greater percentage overlaps than the highest percentage overlap scores of nonstriped species. In these cases of a "perfect" fit (i.e., separation), the maximum likelihood estimate does not exist, so traditional logistic regression and phylogenetically corrected tests are impossible (Albert and Anderson

1984). In contrast, we found no significant associations of presence of body stripes with spotted hyenas, woodland, or any maximum temperature from 25°C to 30°C, mean, or maximum group sizes in single factor analyses controlling for phylogeny [8].

At the subspecies level, body stripe presence (*E. b. quagga* scored as yes) was again "perfectly" associated with presence of tsetse fly distribution, such that we could not run the statistical test. In contrast, we found no significant associations with seven consecutive months of tabanid activity, hyena, woodland, or lion distributions, any maximum temperature, mean, or maximum group sizes in single factor analyses [9].

8.2.c Striping on different parts of the body

We next examined different parts of the body separately because striping may be found only in certain regions (e.g., legs and body of African wild ass and fore- and hindquarters of the extinct *E. b. quagga*). (a) Number of face stripes was associated with six months of consecutive tabanid activity at the subspecies level (Table 8.1) and marginally associated with nine months of consecutive tabanid activity at the species level (Table 8.2). (b) Number of neck stripes was associated with six months of consecutive tabanid activity (Table 8.1) for subspecies, and marginally with nine months of activity for species (Table 8.2). (c) Flank striping was strongly associated with eight months of tabanid activity for subspecies (Table 8.1) and with no factor significant at the species level (Table 8.2). (d) At the subspecies level, rump striping was significantly associated with nine months of tabanid activity (Table 8.1), and marginally with tsetse distribution (Table 8.1). At the species level, rump striping was significantly associated with hyena distribution (Table 8.2). (e) Severity of shadow striping was strongly associated with six months of tabanid activity at the subspecies level (Table 8.1), but we were unable to run a multifactorial logistic regression at the species level because shadow stripes are only found in the plains zebra, although this species does have the highest tabanid activity score of any equid (i.e., presence of shadow stripes is "perfectly" associated with tabanid activity). (f) Belly stripe number was strongly associated with tsetse fly distribution for subspecies (Table 8.1). For species, it was strongly associated with tsetse fly distribution (Table 8.2) and marginally with woodland (Table 8.2). (g) Leg stripe density was significantly associated with seven months of tabanid activity (Table 8.1) for subspecies (Plate 8.1). Leg stripe density was associated with hyena distribution (Table 8.2) and marginally with seven months of tabanid activity at the species level (Table 8.2).

Table 8.1. Full phylogenetic generalized least squares models showing associations between measures of striping and environmental factors at the subspecies level (N = 20).

Variable	z-value	p-value	Importance score
Number of face stripes			
Intercept	0.640	0.522	
6-months tabanids	2.522	**0.012**	0.87
Spotted hyena	0.363	0.716	0.18
Lion	0.093	0.926	0.17
Tsetse flies	0.245	0.806	0.16
Woodland	0.209	0.835	0.16
Temperature 27°C	0.162	0.871	0.16
Number of neck stripes			
Intercept	0.846	0.397	
6-months tabanids	1.961	**0.050**	0.67
Tsetse flies	0.835	0.404	0.24
Lion	0.448	0.654	0.19
Spotted hyena	0.329	0.743	0.18
Woodland	0.060	0.952	0.16
Temperature 27°C	0.007	0.994	0.16
Flank striping			
Intercept	0.433	0.665	
8-months tabanids	2.209	**0.027**	0.77
Spotted hyena	1.124	0.261	0.31
Temperature 28°C	0.453	0.651	0.20
Woodland	0.361	0.718	0.18
Lion	0.122	0.903	0.17
Tsetse flies	0.192	0.848	0.17
Rump striping			
Intercept	0.230	0.818	
9-months tabanids	1.974	**0.048**	0.67
Tsetse flies	1.859	0.063	0.60
Woodland	1.051	0.293	0.28
Spotted hyena	0.610	0.542	0.20
Temperature 30°C	0.696	0.487	0.19
Lion	0.327	0.744	0.18
Shadow stripe severity			
Intercept	0.795	0.427	
6-months tabanids	2.515	**0.012**	0.87
Woodland	0.374	0.709	0.18
Spotted hyena	0.071	0.943	0.17
Tsetse flies	0.187	0.852	0.17
Lion	0.307	0.759	0.16
Temperature 25°C	0.160	0.873	0.16

Table 8.1. Continued

Variable	z-value	p-value	Importance score
		Belly stripe number	
Intercept	0.111	0.912	
Tsetse flies	5.034	**5 × 10⁻⁷**	1.00
12-months tabanids	0.805	0.421	0.22
Woodland	0.738	0.461	0.21
Spotted hyena	0.293	0.779	0.16
Temperature 26°C	0.144	0.885	0.15
Lion	0.072	0.943	0.15
		Leg stripe intensity	
Intercept	0.137	0.891	
7-months tabanids	2.726	**0.006**	0.87
Spotted hyena	1.400	0.161	0.56
Tsetse flies	1.269	0.205	0.33
Temperature 27°C	0.731	0.465	0.26
Lion	0.464	0.643	0.23
Woodland	0.392	0.695	0.16

Note: Significant results in bold.

8.2.d Evaluating the hypotheses

Our comparative analyses that pitted five of the principal hypotheses for striping in zebras against one another indicate that biting flies are responsible for the maintenance of stripes in equids on most regions of the body (Tables 8.1 and 8.2). Many of our independent variables were highly intercorrelated; for instance, spotted hyenas inhabit the hot tropics and subtropics (Kruuk 1972), and tsetse flies rest on trees (Challier 1982). Therefore, we can only be confident of findings stemming from a model selection procedure that pits factors against one another to see the relative importance of each in explaining variation across all possible models.

Biting flies

Our significant findings were (i) that presence of body stripes is associated with tsetse fly distribution and 5–11 consecutive months of tabanid activity at the species level, and with tsetse fly distribution at the subspecies level; (ii) that aspects of face, neck, flank, and rump striping and shadow stripes are associated with between 6 and 9 consecutive months of tabanid activity at the subspecies level; and (iii) that belly stripe number is associated with tsetse fly distribution at the subspecies level, and with tsetse flies

Table 8.2. Full phylogenetic generalized least squares models showing associations between measures of striping and environmental factors at the species level (N = 7).

Variable	z-value	p-value	Importance score
	Number of face stripes		
Intercept	0.606	0.544	
9-months tabanids	1.938	0.053	0.29
Lion	1.459	0.145	0.11
Temperature 25°C	1.014	0.311	0.05
Tsetse flies	0.876	0.381	0.04
	Number of neck stripes		
Intercept	0.613	0.540	
9-months tabanids	1.909	0.056	0.28
Lion	1.451	0.147	0.12
Temperature 25°C	0.952	0.341	0.04
Tsetse flies	0.791	0.429	0.03
	Flank striping		
Intercept	0.670	0.503	
8-months tabanids	1.385	0.166	0.12
Lion	1.347	0.178	0.12
Temperature 25°C	0.968	0.333	0.06
Average group size	0.621	0.535	0.03
	Rump striping		
Intercept	0.268	0.789	
Spotted hyena	2.592	**0.010**	0.60
Temperature 26°C	1.330	0.184	0.09
7-months tabanids	0.938	0.348	0.03
Tsetse flies	1.043	0.297	0.03
	Belly stripe number		
Intercept	0.038	0.970	
Tsetse flies	2.862	**0.004**	0.66
Woodland	1.815	0.070	0.11
11-months tabanids	1.336	0.182	0.04
Temperature 30°C	1.154	0.249	0.03
	Leg stripe intensity		
Intercept	0.454	0.650	
Spotted hyena	2.195	**0.028**	0.36
7-months tabanids	1.820	0.069	0.19
Temperature 25°C	1.372	0.170	0.07
Average group size	1.234	0.217	0.05

Note: Significant results in bold.

and marginally with woodlands at the species level. Indeed, the ranges of where tabanids and tsetse flies are active match the ranges of striped equids well (Plate 8.2). We had estimated 15°–30°C and 30–85% humidity as ideal conditions for tabanid activity and reproduction (see section 8.2.b) but recognize that Culicidae (mosquitoes) and *Stomoxys* (Muscidae, stable flies) and Simuliidae (blackflies) are abundant in the wet season too (G. Gibson and Torr 1999). While the association between striping and warm, humid months might imply stripes being a means of reducing heat load, this seems unlikely given the lack of association with any maximum temperature in any multivariate model.

Instead, our comparative findings lent strong ecological and comparative support to experiments that show that some tsetse and tabanid species avoid black and white striped surfaces (section 5.1). Biting flies are attracted to hosts by a combination of odor, temperature, vision, and movement, which, as mentioned, may act at different stages during host seeking (Vale 1974; Green 1986; Torr 1989; Zollner et al. 2004), but vision is thought to be important in the landing response from about 5–20 m away from the target, depending on the species (G. Gibson and Torr 1999), with dark colors being particularly attractive (Allan, Day, and Edman 1987; Green 1986; Holloway and Phelps 1991; Steverding and Troscianko 2004). Following Egri, Blaho, Kriska, and colleagues (2012), from the literature we recalculated glossinid, tabanid, and *Stomoxys* preferences for varying stripe widths. We measured widths of both black and white stripes along "transects" on the cheek (muzzle to a line perpendicular to the ear), neck (ear to the midpoint between shoulder and forelimb), side (middle of body from forelimb to hindlimb), rump (diagonal line beginning at a point on the spine above the hindlimb stretching to lower edge of the rump for mountain and plains zebra, but from the dorsal base of the tail to center of rump and then vertically down the same distance to form a Y shape in Grevy's zebra), foreleg, and hindleg (Plate 8.3). Data obtained from these different transects were first averaged across individuals of each species, and then an average of each region of the body was calculated for the three zebra species.

Zebra stripe widths on various parts of the body fall in the range lower than that eschewed by all three groups of biting flies (Figure 8.1), extending Egri, Blaho, Kriska, and coworkers' (2012) analysis to two other families of biting flies. It is noteworthy that stripes nearer the ground (on the legs and head when grazing) are especially thin and biting flies prefer to alight on shaded parts of animals low on the body (Allan, Day, and Edman 1987). Specifically, tabanids cruise at around 30 cm above ground (Kirk-Spriggs 2012) and prefer alighting on lower legs of cows (Mohamed-Ahmed and Mihok 2009), although the exact area differs by tabanid species (Mullens

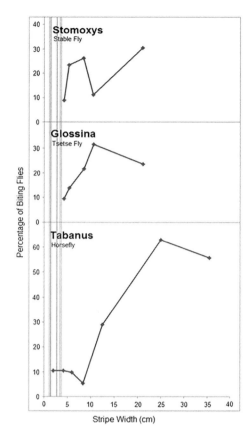

Figure 8.1. Biting fly preferences and stripe widths. Percent of biting flies landing on stripes of varying widths. *Stomoxys* and *Glossina* from Brady and Shereni (1988), Tabanidae from Egri and colleagues (2012). Vertical lines show average stripe widths taken from (left to right) face (1.21 cm), legs (1.35 cm), neck (3.17 cm), flank (3.74 cm) and rump (3.90 cm) averaged across the three species of zebra (from Caro et al. 2014).

and Gerhardt 1979; Phelps and Holloway 1990); stable flies concentrate low on cows' bodies (Ralley, Galloway, and Crow 1993), as do blackflies (J. Anderson and DeFoliart 1961); and tsetse flies prefer to feed on the hocks of cattle (Kangwagye 1976; M. Thompson 1987; Torr and Hargrove 1998). Interestingly, leg striping is seen in the African wild ass, and faint leg striping is sometimes seen on some Przewalski's horse individuals, although on none in our sample.

Is there evidence that zebra striping deters biting flies from landing? Indirect observations are suggestive. First, incidence of tsetse blood meals derived from zebras is low in the wild (Clausen et al. 1998), particularly in *Glossina morsitans* (Weitz 1963; Glasgow 1967). Second, the incidence of trypanosomiasis in zebras (carried by tsetse flies and horseflies) is either lower (Harrison 1940; Willett 1960–1961) or comparable with sympatric mammals (Vanderplank 1942; de Andrade Silva and Marques da Silva 1959), whereas domestic horses (unstriped) suffer badly from this disease in many parts of Africa (Dorward and Payne 1975; Auty et al. 2008; Pinchbeck et al. 2008). Elsewhere feral and semidomesticated horses can be plagued by flies (Rutberg 1987; Rubenstein and Hohmann 1989). Many herbivores are subject to intense fly annoyance that can reduce time spent feeding (Toupin, Hout, and Manseau 1996) yet have a suite of strategies to reduce attack, including moving to different habitats (Downes, Therberge, and Smith 1986), resting (Espmark and Langvatn 1979), bunching together (Ralley, Galloway, and Crow 1993), or retreating to shade (Keiper and Berger 1982). Moreover, feral horses employ the same behaviors of moving to safer areas to escape flies (Duncan and Cowtan 1980; Keiper and Berger 1982; Rubenstein and Hohmann 1989; see also King and Gurnell 2010), moving into shade (Horvath et al. 2010), or grouping together (Duncan and Vigne 1970) to ameliorate biting fly attack.

Why therefore should African equids be so sensitive to biting flies and have evolved morphological as well as behavioral defenses? Hannah Walker scored hair depths and lengths on museum pelts of plains zebra, Grevy's zebra, mountain zebra, Asiatic wild ass, and Przewalski's horse, as well as artiodactyls that live sympatrically with zebra species: blue wildebeest, roan antelope, oryx, hartebeest, waterbuck, and topi. She placed a millimeter metal ruler on nine places on museum pelts of these species (Pan 1964)—forehead, neck, side, belly, midrump, and upper and lower forelegs and hindlegs—and measured the depth of fur perpendicular to the skin (Moen and Severinghaus 1984). She measured the length of the longest hair shafts at eight sites: neck, side, belly, buttock, and upper and lower hindlimbs. We also extracted lengths of biting fly mouthparts from the literature (Jobling, n.d.; Pisciotti and Miranda 2005; Krenn and Horst 2012).

Hair depths and hair lengths of zebras are shorter than sympatric African artiodactyls and Przewalski's horse (Figure 8.2), potentially making them more susceptible to biting flies probing perpendicularly or burrowing into and against the hair shafts. Alternatively, stripes are a sufficient deterrent to allow zebras to have cropped pelage. Moreover, unlike Asiatic equids, zebras do not have thick winter coats. In cattle, landing area hair depth and tabanid labrum lengths are positively correlated (Mullens and Gerhardt 1979), yet in zebras biting fly mouthpart lengths are markedly longer than average hair depths and approximately the same as zebra hair lengths (Figure 8.2). This may make zebras particularly susceptible to annoyance everywhere on their bodies.

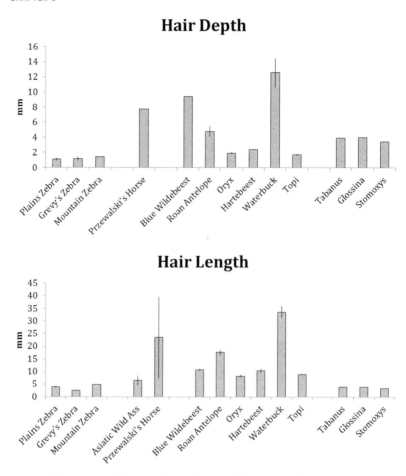

Figure 8.2. Pelage characteristics of herbivores. Mean and SE hair depths (*above*) and mean and SE longest hair lengths (*below*) of zebra species, nonstriped equids, and artiodactyls living sympatrically with zebras measured using standard techniques. Plains zebra (5, 14 pelts, respectively), Grevy's zebra (3, 6), mountain zebra (1, 1), Asiatic wild ass (0, 7), Przewalski's horse (1, 2), blue wildebeest (1, 8), roan antelope (4, 18), oryx (3, 9), hartebeest (1, 7), waterbuck (3, 9), and topi (7, 13). Lengths and depths calculated from taking the averages at seven points on the body (neck, flank, belly, rump, and upper and lower foreleg and hindleg) for each pelt and averaging across pelts. Also shown are lengths of mouthparts of biting flies (from Caro et al. 2014).

At an ultimate level, blood loss from biting flies can be considerable. Calculations show that blood loss from tabanids alone can reach 200–300 cc/cow/day in the United States (Tashiro and Schwardt 1949; Hollander and Wright 1980). For example, in Pennsylvania, mean weight gain per cow was 37.2 lb lower over an 8-week period in the absence of insecticide that prevents horn fly (*Siphona*), stable fly (*Stomoxys*), and horsefly

(*Tabanus*) attack (Cheng 1958), and in New Jersey, milk production increased by 35.5 lb over a 5-week period when insecticide was applied to cows (Grannett and Hansens 1956). Milk loss to stable flies was calculated at 139 kg/cow/annum for the United States (Taylor, Moon, and Mark 2012). Similarly, bloodsucking insects have been shown to negatively impact performance in draft horses. In a 1981 study conducted in a forest in the Czech Republic, applying fly repellant to a test group of horses increased their performance in wood-skidding operations by 58% compared to a control group receiving no repellant (Riha et al. 1983).

Alternatively or additionally, striped equids might be particularly susceptible to certain diseases that are carried by fly vectors in Sub-Saharan Africa (Krinsky 1976; Foil 1989). Hannah Walker collated literature on diseases carried by biting flies that attack equids in Africa (Table 8.3) and found that trypanosomiasis, equine infectious anemia, African horse sickness, and equine influenza are restricted to equids, all are fatal, and all are carried by tabanids. Currently, it is difficult to disentangle whether zebras are particularly vulnerable or susceptible to biting flies because they carry dangerous African diseases or because of excessive blood loss, but I am inclined toward the former because Eurasian equids are not striped yet are demonstrably subject to biting fly annoyance (Horvath et al. 2010).

Background matching

As important as uncovering positive associations between striping and biting fly annoyance is the failure to find positive associations between striping and other factors. We found no comparative evidence for stripes being associated with woodland habitats, which argues against striping being a form of background matching in this environment. For subspecies, there was a marginal association between number of belly stripes and living in woodland habitat, but this likely derives from the close association between tsetse flies and woodlands rather than striping being a form of crypsis. Indeed, we found no evidence of striping elsewhere on the body being associated with woodland habitat, and it is difficult to envisage belly striping as a form of background matching given its ventral positioning. Many artiodactyls that have light or white stripes and spots set on a light or dark brown coat live in lightly wooded forests (Stoner, Caro, and Graham 2003). Equids, on the other hand, are an open-country clade with asses inhabiting deserts, zebras inhabiting tropical grasslands, and horses living in temperate grasslands (Table 1.2), and background matching against grassland has been discounted by other authors (Godfrey, Lythgoe, and Rumball 1987).

Table 8.3. Diseases infecting equids that are transmitted by biting flies in Africa.

Disease name	Transmitted by					Known in zebra?	Disease effects in equids	Disease effects in artiodactyls
	Horsefly	Tsetse fly	Stable fly	Mosquito	Midge			
Trypanosomosis: Nagana	Yes[1]	Yes[2]				Yes[3]	Fever, weakness, lethargy, weight loss; eventual death in all domestics unless treated.[3]	Cattle, sheep, goats: similar symptoms.[3] Many wild ungulates native to Africa show no evidence of harm.[1]
Equine infectious anemia	Yes[4]		Yes[4]			Yes[5]	Anemia, swelling, death, fetal abortion.[6]	Not known outside of equids.[1]
African horse sickness*	Yes[7]				Yes[8]	Yes[1]	Fever followed by dyspnea and spasmodic coughing. Death within a week, recovery rare.[1] Zebra often experience inallarent form of disease with prolonged viremia.[9]	Disease unknown outside of equids,[10] although antibodies detected in other animals.[1]
Anaplasmosis	Yes[11]					Yes[2]	Inappetence, loss of coordination, breathlessness when exerted, and a rapid, bounding pulse.[7]	Cattle, sheep, goats, buffalo, and some wild ruminants: similar symptoms.[1]
Anthrax	Yes[6]					Yes[1]	Fever, sudden death.[9]	Cattle and sheep: same effects.[1]
Babesiosis	Yes[7]					Yes[7]	Persistent fever, inappetence, weight loss.[1]	Cattle, sheep, goats (wide range): similar symptoms.[1]
Equine influenza	Yes[2]					Yes[1]	Respiratory infection, fatal particularly in zebras.[1]	Not known outside of equids.[1]

Disease				Clinical signs	Other hosts
Eastern equine encephalitis	Yes[1]		Yes[12]	Paralysis, seizures, and death. Mortality 90%.[1]	Other arboviruses cause diseases in other animals. Those that cause EEE are transmitted by mosquitoes and birds, and can infect other mammals.[1]
Equine encephalosis*	Yes[7]		Yes[13]	One to five days of fever, inappetence, sometimes females abort. Infects all equids. Clinical signs only seen in horses.[13]	Not known outside of equids, although some evidence elephants may become infected.[13]
Trypanosomosis: Dourine*	Yes[7]		Likely[2]	Paralysis, progressive emaciation. Fatality 50%–70%.[1]	Not known outside of equids.[2]
Trypanosomosis: Surra	Yes[14]	Yes[10]	Not known	Fever, weakness, lethargy, weight loss; fatal when not treated.[15]	All domestic animals susceptible, similar symptoms.[1]

*Disease found only in Africa.
[1]Aiello (2002–2012).
[2]Foil (1989).
[3]Spinage (2012).
[4]Factsheets: Equine Infectious Anemia (2009).
[5]Yapic et al. (2007).
[6]Muoira et al. (2007).
[7]Radcliffe and Osofsky (2002).
[8]Krinsky (1976).
[9]Brown (1992).
[10]Lehane (2005).
[11]Hornok et al. (2008).
[12]Factsheets: Equine Encephalitis (2009).
[13]Wohlfender (2009).
[14]Lane and Crosskey (1993).
[15]World Organization for Animal Health (OIE) (2009).

Predation

Hypotheses for body striping in equids that center on avoiding predation through various forms of predator confusion (Eltringham 1979;
Cloudsley-Thompson 1984; Kingdon 1984; Stevens et al. 2011; How and
Zanker 2013) all tacitly demand that stripes should have evolved in the
presence of large predators. Yet there was no strong congruence between
striping on any part of the animal at the species or subspecies level and
close sympatry with lions, except for leg stripe intensity at the species
level, driven particularly by the extinct Barbary lion and extinct Atlas African wild ass co-occurring in the Maghreb. As mentioned in section 4.10,
across contemporary ecosystems in Africa where lions have been subjects
of repeated detailed study, lions capture zebras in significantly greater proportions to their abundance (Hayward and Kerley 2005), suggesting that
stripes are not an antipredator defense against lions acting through confusion, motion dazzle, or other means. Indeed, lions and other African
carnivores do not usually make a final leap at or ahead of their quarry but
knock, climb, or savage it following a short chase (Ruxton 2002). Despite
the popularity of various sorts of confusion hypotheses, our comparative
data provided little support for this idea, adding further refutation to the
observations and experiments described in Chapter 4.

In contrast, spotted hyenas, which are smaller than lions, capture fewer
zebras than predicted by their relative abundance (Hayward 2006). Hyenas find it difficult to pull down an adult zebra because zebras can deliver sharp kicks and bites to these predators; instead, hyenas concentrate
mostly on foals (Kruuk 1972). We found significant associations with rump
striping and leg stripe intensity at the species level and nonsignificant but
reasonably high importance values between flank and rump striping and
leg stripe intensity at the subspecies level and overlap with the geographic
range of spotted hyenas (I = 0.31, 0.20, 0.56, respectively; see Table 8.1).
Confusion seems improbable, as spotted hyenas and other coursing predators do not leap at their quarry but bite at its hindlegs or belly when making first contact. Instead, the possibility that striping on these areas of the
body (where coursing predators might have prolonged views) is an aposematic signal to warn spotted hyenas (and perhaps even smaller leopards
and cheetahs, which also eschew zebras; Hayward, Henschel, et al. 2006;
Hayward, Hofmeyr, et al. 2006) to keep away cannot be dismissed entirely.
Nevertheless, there were no significant species associations between hyena
distribution and striping measures on other areas of the body, and none at
the subspecific level: aposematic deterrence of hyenas could possibly be a
secondary benefit of stripes but cannot explain the presence of stripes on

zebras more generally. In parts of Asia once inhabited by tigers and wolves, equids are not striped.

Intraspecific communication

We found no significant comparative associations between any aspect of striping and group size, although mean and maximum group sizes are of course crude measures of the relative intensity of social interactions that have received uneven study across equid species. Given equids' two sorts of social system (Klingel 1974; Rubenstein 1986), a single stallion that lives with a small harem of mares and nonharem holding stallions that form small bands (Type I), or loose bands of animals that collect together in medium-size to large aggregations at certain times of the year (Type II), we might expect individual recognition based on stripe pattern to be more advantageous in the Type II system. But three out of these four species do not have body stripes. As described in section 6.4, domestic horses are capable of recognizing other individuals (Kimura 1998; Sigurjonsdottir et al. 2003) using visual, auditory, and olfactory cues (Rubenstein and Hack 1992; Proops, McComb, and Reby 2009) and do not have stripes, so it seems unlikely that stripes are a critical means of identification.

Cooling

Across species there were no significant associations between striping at species or subspecies levels and inhabiting hot areas using six maximum temperature measurements. In fact, all equids live in open areas, and midday temperatures reach high figures in parts of the geographic ranges inhabited, for example, by plains and Grevy's zebras (east and southern Africa, northeast Africa, respectively) and unstriped African and Asiatic wild asses (northeast Africa and central southern Asia, respectively) (Bauer, McMorrow, and Yalden 1994). Both striped and unstriped equids use behavioral means to manage heat stress by seeking available shade, suggesting that striped species have no substantial advantage in this domain.

8.3 Conclusions

We matched variation in striping of all seven equid species and striping of eighteen extant and two extinct subspecies, to geographic range overlap of environmental and social variables in multifactor models controlling for phylogeny in order to test five major explanations for patterns of zebra striping simultaneously. For subspecies, there were significant associations between our environmental proxy for tabanid biting fly annoyance and most measures of striping (e.g., facial and neck stripe number, flank and rump striping, leg stripe intensity, shadow striping), and between belly stripe number and tsetse fly distribution, several of which were replicated at the species level. Conversely, there was no consistent support for camouflage, predator avoidance, social interaction, or cooling hypotheses. We further examined susceptibility to ectoparasite attack in relation to short coat hair, disease transmission, and blood loss.

These comparative phylogenetic analyses titrate hypotheses for the functions of striping in equids and lend strong ecological support for striping being an adaptation to avoid biting flies. Nonetheless, caution is warranted. First, our attempts to extract variables to test each hypothesis were inexact: plotting the distribution of tabanids was very difficult, as there are no range maps for the more than 4000 species, and estimates of abundance are restricted to isolated study sites, often in the New World. It is hoped that future studies can generate better quantitative spatial measures of tabanid annoyance. We were confident about historical tsetse fly distributions, historical predator ranges, and annual temperatures, however. A second issue is that comparative analyses suffer from an inescapable lack of statistical power because there are only seven extant species and twenty-three subspecies for which we have coarse environmental information, three of which had to be dropped due to inadequate resolution of our phylogenetic tree. Furthermore, tree reconstruction at the species level shows that striping evolved only once in this clade, and further studies may alter the phylogenetic tree itself. Third, it is difficult to separate the effects of striping per se from the effects of having two-tone pelage. Put another way, are the ecological associations with stripes or with having black and white pelage of any pattern juxtaposed? That said, it is reassuring that the sequence of studies indicating that biting flies shun striped surfaces, first put forward by R. Harris in 1930, is given ecological validity by our finding that striping on equids is perfectly associated with increased presence of biting flies.

8.4 The intraspecific comparison

Focusing on plains zebras, the most variable species in regard to striping patterns, Larison, Harrigan, Thomassen, and colleagues (2015) conducted an investigation into the factors that explain intraspecific geographic variation in striping. They visited sixteen populations of zebras and photographed a minimum of eight individuals in each. For every zebra they quantified the number, length, stripe thickness, and saturation of stripes on the torso, legs, and belly. Subsequently they derived an additional measure of stripe definition (standardized values of thickness, length, and saturation multiplied together).

Then they sought out a number of environmental variables, which included temperature and precipitation measures from WorldClim (http://www.worldclim.org/) and remote sensing variables quantifying the concentration of green leaf vegetation (MODIS-derived Normalized Difference Vegetation Index [NDVI]) and tree canopy cover. They also examined surface water, leaf water content and seasonal attributes from Quick Scatterometer (QSCAT), FAO data on tsetse fly occurrence, and data on lion distribution from MAXTENT; the latter two distributions were based on habitat suitability modeling. Random forest models were used to match stripe traits to a total of nineteen environmental variables.

Environmental variables explained 30–63% of the variance in twelve of eighteen striping measures: foreleg stripe number, thickness, saturation, and overall definition; hindleg stripe thickness and definition; torso stripe number, length, thickness, saturation, and definition; and belly stripe number. The most consistently important environmental variables were isothermality ([mean of the monthly {maximum daily temperature – minimum daily temperature}] / [maximum temperature of the warmest month – minimum temperature of the coldest month]) (BIO3 from WorldClim), the mean temperature of the coldest quarter (BIO11), maximum annual vegetation (NDVIMAX), and precipitation of the wettest month (BIO13). To illustrate, distinct saturated striping is found in the warmer, northerly part of the plains zebra's range, but less distinct stripes are found in cooler southern Africa. Estimated tsetse fly and lion distributions failed to predict variation in striping.

They then used random forest models to predict striping at eight additional test sites, even though their number of zebra populations was low, testing the ability of constant temperatures, mean temperatures during the coldest quarter of the year, and precipitation to predict striping attributes intraspecifically. The validity of model estimates was well supported for foreleg stripe length, thickness, saturation, and definition; hindleg stripe thickness and definition; and number of torso stripes and their definition; but not for belly stripe measures.

The difficulty lies in interpreting the results (Caro and Stankowich 2015; Larison, Harrigan, Rubenstein, and Smith 2015). The first point of explanation, perhaps because it was suggested a quarter of a century ago (Morris 1990), might lie in well-defined saturated stripes setting up convection currents due to black pelage absorbing but white pelage reflecting thermal radiation, and Larison and coworkers suggested this might be operating on the torso. There are concerns, however: black pelage has a higher heat load than other pelage types and so would be expected to be reduced in hotter northerly parts of the continent; shadow stripes found in some central and southerly populations could disrupt the reflective properties of white stripes; and, importantly, convection currents would be ineffective in wind or when zebras walk. Moreover, this account does not explain why statistical models explaining torso stripe number and definition in the test data set included high precipitation as well as isothermality. Another explanation is that the white stripes on plains zebras in eastern Africa are a bright, conspicuous white and reflect sunlight well, whereas the white stripes on plains zebras from southern Africa do not have to reflect solar radiation so effectively and consequently are creamy white or yellow in hue (D. Rubenstein, personal communication). This second possibility does not demand convection currents.

A different explanation also raised by Larison and her team is that stripes on the forelegs and hindlegs are a response to biting flies. Given that they found no association between striping and tsetse fly occurrence, they suggest instead that trypanosome infections are temperature dependent (see Welburn and Maudlin 1991). This is an interesting idea because their temperature and precipitation variables may have more relevance to disease prevalence such as trypanosomiasis than to their glossinid vectors. Their intraspecific study did not examine tabanid occurrence, however, vectors that also carry infectious diseases (Table 8.3), and it may be that isothermality, warm temperatures, and strong precipitation exert their effect on striping via increased tabanid fly annoyance, or disease prevalence, or both.

8.5 Concordance on multifactorial analyses

Superficially, these two multifactorial analyses give disparate results, but they actually agree on many issues (Caro and Stankowich 2015) in spite of being conducted at different taxonomic levels (Table 8.4). Neither study found evidence for striping being associated with woodlands or tree cover. Neither study found convincing evidence for the geographic distribution of stripes on any part of the body being associated with the distribution of lions in Africa. Neither study found overwhelming evidence for tsetse fly

Table 8.4. Summary of findings from the two multifactorial studies of striping in zebras.

Test: Hypothesis	Comparative Caro et al. (2014)	Intraspecific Larison, Harrigan, Thomassen, et al. (2015)	Notes
Predation			
Crypsis	No	No	No association with trees
Lion	No	No	Zebras are preferred prey
Spotted hyena	Unlikely	Not tested	
Social interaction	No	Not tested	All equids highly social
Antiparasite			
Glossinid	No and yes (belly)	No	Carry diseases fatal to equids and do not like to land on stripes
Tabanid	Yes	Not tested	Carry diseases fatal to equids and do not like to land on stripes
Cooling	No	Yes	Mechanism unclear: includes both temperature regulation and disease

Source: From Caro and Stankowich (2015).

distributions being related to striping on different parts of the body; the only significant association was between belly stripe number and tsetse flies in the comparative analysis, although this was highly significant. Interestingly, belly stripe characteristics are not associated with stripe characteristics on other areas of the body (Larison, Harrigan, Thomassen, et al. 2015), possibly suggesting that different selection pressures are acting in this region of the torso.

The studies differ in that we found no association between measures of striping and mean annual maximum temperatures between 25°C and 30°C (Tables 8.1 and 8.2), whereas Larison and her colleagues uncovered positive associations with isothermality and mean temperature in the coldest quarter of the year. Yet our proxy measure of tabanid abundance was temperatures lying between 15°C and 30°C and humidity between 30% and 85% for six to seven months or more of the year. Such warm, humid conditions are highly congruent with Larison and coworkers' precipitation during the wettest month of the year (BIO13) and constant temperatures (BIO3), which were associated with striping in plains zebras. Thus, it is quite probable that these two WorldClim measures along with warmer temperatures during cold seasons (BIO11) could be proxies for tabanid activity too, or at least proxies for infection rates that adversely affect zebras.

In conclusion, the two multifactorial analyses have concordance in dismissing lion predation as a driver of striping, failing to find an association between striping and woodland habitat, casting doubt on tsetse flies being of major importance for striping, and agreeing that stripes cannot generate cooling eddies on the legs (Larison, Harrigan, Thomassen, et al. 2015) or anywhere on the body (Caro et al. 2014). Instead, both studies point indirectly, by way of proxy environmental variables, to tabanid or other biting fly annoyance as the evolutionary driver of striping in Equidae, probably because of the diseases that they transmit. I develop these ideas in the next chapter.

Statistical tests

[1] Face stripe number. Subspecies level: single factor analyses (PGLS: N = 20) produced p-values as follows: maximum temperature of 27°C (p = 0.609), 6 consecutive months of tabanid activity (p = 0.009), tsetse fly distribution (p = 0.723), woodland (p = 0.423), hyena (p = 0.265), lion (p = 0.307). Nonsignificant figures shown throughout to highlight lack of support for hypotheses.

Species level: single factor analyses (PGLS: N = 7) produced p-values as follows: maximum temperature of 25°C (p = 0.229), 9 consecutive months of tabanid activity (p = 0.051), tsetse fly distribution (p = 0.027), woodland (p = 0.599), hyena (p = 0.171), lion (p = 0.107), mean group size (p = 0.457), maximum group size (p = 0.386).

[2] Neck stripe number. Subspecies level: single factor analyses (PGLS: N = 20) produced p-values as follows: maximum temperature of 27°C (p = 0.774), 6 consecutive months of tabanid activity (p = 0.040), tsetse fly distribution (p = 0.609), woodland (p = 0.677), hyena (p = 0.503), lion (p = 0.318).

Species level: single factor analyses (PGLS: N = 7) produced p-values as follows: maximum temperature of 25°C (p = 0.253), 9 consecutive months of tabanid activity (p = 0.053), tsetse fly distribution (p = 0.3410), woodland (p = 0.602), hyena (p = 0.196), lion (p = 0.109), mean group size (p = 0.451), maximum group size (p = 0.385).

[3] Flank striping. Subspecies level: single factor analyses (PGLS: N = 20) produced p-values as follows: maximum temperature of 28°C (p = 0.338), 8 consecutive months of tabanid activity (p = 0.0148), tsetse fly distribution (p = 0.284), woodland (p = 0.823), hyena (p = 0.068), lion (p = 0.480).

Species level: single factor analyses (PGLS: N = 7) produced p-values as follows: maximum temperature of 25°C (p = 0.253), 8 consecutive months of tabanid activity (p = 0.126), tsetse fly distribution (p = 0.637), woodland (p = 0.969), hyena (p = 0.173), lion (p = 0.133), mean group size (p = 0.427), maximum group size (p = 0.479).

[4] Rump striping. Subspecies level: single factor analyses (PGLS: N = 20) produced p-values as follows: maximum temperature of 30°C (p = 0.270), 9 consecutive months of tabanid activity (p = 0.011), tsetse fly distribution (p = 0.010), woodland (p = 0.866), hyena (p = 0.069), lion (p = 0.658), mean group size (p = 0.427), maximum group size (p = 0.479).

Species level: single factor analyses (PGLS: N = 7) produced p-values as fol-

lows: maximum temperature of 26°C (p = 0.075), 7 consecutive months of tabanid activity (p = 0.211), tsetse fly distribution (p = 0.217), woodland (p = 0.755),
hyena (p = 0.017), lion (p = 0.229), mean group size (p = 0.601), maximum
group size (p = 0.648).

[5] Shadow stripe severity. Subspecies level: single factor analyses (PGLS: N = 20)
produced p-values as follows: maximum temperature of 25°C (p = 0.777), 6
consecutive months of tabanid activity (p = 0.010), tsetse fly distribution (p =
0.510), woodland (p = 0.756), hyena (p = 0.405), lion (p = 0.262).

Species level: single factor phylogenetic logistic regressions (N = 7) at the 0/1
level produced p-values as follows: maximum temperature of 25°C (p = 0.595),
tsetse fly distribution, woodland, and 3–11 consecutive months of tabanid activity all showed a perfect fit with the presence of shadow stripes, hyena (p =
0.578), lion (p = 0.910), mean group size (p = 0.921), maximum group size (p =
0.633), but we could not run a multivariate model.

[6] Belly stripe number. Subspecies level: single factor analyses (PGLS: N = 20)
produced p-values as follows: maximum temperature of 26°C (p = 0.258), 12
consecutive months of tabanid activity (p < 0.0001), tsetse fly distribution (p <
0.0001), woodland (p = 0.207), hyena (p = 0.032), lion (p = 0.880).

Species level: single factor analyses (PGLS: N = 7) produced p-values as follows: maximum temperature of 30°C (p = 0.468), 11 consecutive months of
tabanid activity (p = 0.130), tsetse fly distribution (p = 0.014), woodland (p =
0.056), hyena (p = 0.657), lion (p = 0.657), mean group size (p = 0.968), maximum group size (p = 0.605).

[7] Leg stripe intensity. Subspecies level: single factor analyses (PGLS: N = 20) produced p-values as follows: maximum temperature of 27°C (p = 0.423), 7 consecutive months of tabanid activity (p = 0.001), tsetse fly distribution (p = 0.031),
woodland (p = 0.598), hyena (p = 0.0083), lion (p = 0.380).

Species level: single factor analyses (PGLS: N = 7) produced p-values as follows: maximum temperature of 25°C (p = 0.120), 7 consecutive months of tabanid activity (p = 0.051), tsetse fly distribution (p = 0.880), woodland (p = 0.909),
hyena (p = 0.098), lion (p = 0.031), mean group size (p = 0.149), maximum
group size (p = 0.263).

[8] Species level. Presence of body stripes with spotted hyenas. Phylogenetic logistic
regression: N = 7 species, p = 0.403), woodland (p = 0.788), or any maximum
temperature from 25°C to 30°C (p = 0.432–0.803), mean (p = 0.951) or maximum group sizes (p = 0.956) in single factor analyses controlling for phylogeny.

[9] Subspecies level. Presence of body stripes with 7 consecutive months of tabanid
activity. Phylogenetic logistic regression: N = 20 subspecies, p = 0.435), hyena
(p = 0.818), woodland (p = 0.926), or lion (p = 0.956) distributions, any maximum temperature (p = 0.585–0.745), mean (p = 0.772) or maximum group sizes
(p = 0.600) in single factor analyses controlling for phylogeny.

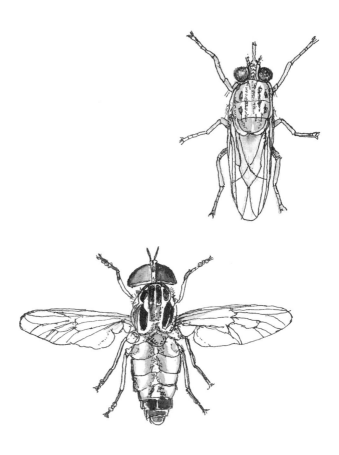

Figure 9.0. *Opposite:* A tsetse fly (*above*) and a tabanid (*below*). These dipterans are thought to drive the evolution of striping in equids. (Drawing by Sheila Girling.)

NINE

The case for biting flies

9.1 Last man standing

Stepping back from discussions about individual hypotheses for the function of zebra stripes, one is left with an overwhelming sense that few of the notions have moved much beyond the "ideas stage," save one. In this chapter I build a case for stripes being an evolutionary response to pressure from biting flies by discussing the causal chain that must be followed in order to link biting fly attack to striping in equids. In the final part of the chapter, I list a series of questions that we need to answer, particularly concerning mechanisms, in order to make our understanding of this unique pattern of external coloration more robust. First, however, I briefly summarize the main reasons that alternative hypotheses for striping are flawed.

There are about eighteen hypotheses for the function of striping in zebras, depending upon one's method of categorization. The arguments for and against each can be summarized as follows (Table 9.1). Antipredation hypotheses seem highly improbable in regard to lions because they kill zebras in greater proportions than expected from their abundance across the continent; so this hypothesis could only apply to spotted hyenas or other large to midsize carnivores. Crypsis seems unlikely because zebras do not skulk, are easy to see in the wild, and their movements and tail swishing are obvious even at a distance. Lions and spotted hyenas can still discern zebras as objects distinct from their backgrounds but cannot resolve stripes at any great distance, so stripes cannot aid in crypsis (Melin et al. 2016). Extent of striping is not associ-

Table 9.1. Summary of findings about the function of zebra striping.

Hypothesis	Findings
Crypsis	
Through background matching in grass or woodland	Behavioral ecology does not conform to that expected of crypsis.
	Zebras easy for humans to see.
	Predators cannot see stripes at a distance.
	Striping not geographically associated with trees.
Through disruptive coloration	Design features do not conform to predictions.
	Zebras easy for humans to see.
	Predators cannot see stripes at a distance.
Through countershading	Not found in predominantly open areas.
	Does not explain striping.
Aposematism	Zebras easy to see close up and noisy.
	Zebras suffer high predation from lions.
Confusion	
Through miscounting	Humans can count zebras.
Through obscuring outline in flight	Zebras flee with gaps in between them.
Through preventing being followed	Zebras do not flee in an unpredictable way.
Through dazzle	No evidence that lions are confused.
Through motion dazzle	Computer experiments provide mixed support for benefits of striping.
	Predators do not make final leaps.
	Zebras' speed not disproportionately difficult to judge.
Through misjudging size	Zebra sizes not disproportionately difficult to judge.
Quality advertisement	Difficult to see how this might evolve.
Predation generally	Lions prefer zebras so could only work against spotted hyenas but more parsimonious to suggest large size, not stripes.
Intraspecific signaling	
Species recognition	Little sympatry with other equids.
Allogrooming	Zebras show little mutual grooming.
Social bonding	No association between striping and close social bonds.
Individual recognition	Nonstriped equids capable of this.
	Striping not geographically associated with group size.
Indicator of quality	No association with fluctuating asymmetry.
Thermoregulation	Improbable on comparative and mechanistic grounds.
	Zebras not cooler than other herbivores.
	Intraspecific association between extent of striping and temperature.
Ectoparasites	Several experimental studies show glossinids, tabanids, and *Stomoxys* avoid landing on stripes.
	Confirmed for tabanids but not strongly for glossinids in this study.
	Comparative data show close association between striping and tabanid annoyance.
	Not confirmed for glossinids intraspecifically.

ated with woodland or tree cover. Stripes might be an aposematic signal directed at spotted hyenas, but the extent to which they make zebras' defenses appear more dangerous than those of other similar-size herbivores is unknown, and under this hypothesis it is unclear why the zebra's whole exterior should be striped. Predator confusion has little merit because zebras are easy to count; zebra flight behavior does not make it difficult to pick out an individual; and stripes do not promote human misjudgments over individuals' speed or size. Laboratory motion dazzle experiments with people generate mixed results in regard to striping and fail to mirror the final stages of predatory behavior of either spotted hyenas or large carnivores. These psychological experiments involving computers are seductive because they are (relatively) easy to carry out and they explore interesting ideas that are tangentially related to both zebra pelage and dazzle coloration of World War I vessels. Yet their relevance to the evolution of zebra striping is questionable because they disregard the natural history of hunting by large mammals and the extent to which zebra populations are limited by predation.

Ideas about stripes being involved in species recognition seem improbable because allopatry predominates (no geographic range overlap). Intraspecific signaling falls short because rates of allogrooming and social interactions that are supposedly enhanced by stripes are not particularly extensive in zebras. Extent of striping is not associated with group size metrics. A third idea, individual recognition, seems improbable because many mammals can recognize one another without recourse to stripes. Indicators of quality seem unlikely on logical grounds.

Stripes being a cooling mechanism is improbable given that a convection mechanism could only operate at limited times of the day, and it relies on a unique solution to heat management in terrestrial vertebrates: other mammals in hot climes are pale rather than striped. My empirical evidence does not show that zebras are cooler than sympatric herbivores. There are no interspecific associations between striping and warm temperatures, but there are intraspecific associations, although heat management is only one of several explanations for this.

Stripes as a means of avoiding biting flies is the only hypothesis that bears promise because it is based on two independent avenues of research: repeated experiments in the field and laboratory, and analyses of striping and fly geographic range overlap across equid species and subspecies. I now flesh out the ectoparasite solution to zebra external coloration and show where it has biological merit and where gaps in our knowledge lie. So rather than using observations, experiments, museum information, or phylogenetic methods, I am turning to the literature.

9.2 Host choice

It is important to establish which groups of biting flies potentially cause the most annoyance to zebras, and while much is known about general hosts of most dipteran ectoparasites, less is known about their species preferences. Sandflies (Phlebotominae) feed on reptiles and homeotherms (Lane 1993); mosquitoes (Culicidae) on terrestrial vertebrates that vary according to species and region (Washino and Tempelis 1983); blackflies (Simuliidae) on birds and mammals (Crosskey 1993a); biting midges (Ceratopogonidae) on vertebrates (Boorman 1993); horseflies, deerflies, and clegs (Tabanidae) on monkeys, equids, bovids, ruminants, rodents, birds, and some reptiles (Chainey 1993); stable flies and horn flies (bloodsucking Muscidae) on livestock (Crosskey 1993b); and tsetse flies (Glossinidae) on a variety of vertebrate hosts, principally Suidae and Bovidae (Jordan 1993). For glossinids that transmit sleeping sickness, there is a body of knowledge about species-specific relationships between parasites and hosts (Harrison 1940). Very briefly, the *Glossina morsitans* group dines on suid and bovid ruminants, depending on the tsetse fly species; the *Glossina palpalis* group feeds on bovids, with *G. palpalis* and *G. fuscipes* focusing on bushbuck, and on domestic animals, principally pigs; whereas the *Glossina fusca* group feeds principally on suids and bovids, with *G. fuscipleuris* concentrating on bushpig and *G. brevipalpis* focusing on hippopotamus (Table 9.2).

Interestingly, there are very few mentions of tsetse flies landing on zebras. Collating a large number of studies that identified tsetse blood meals—nine of *Glossina morsitans morsitans*; seven of *G. palpalis palpalis*; six of *G. swynnertoni*; five of *G. morsitans submorsitans*; four each of *G. brevipalpis* and *G. pallidipes*; three of *G. fusca*; two each of *G. medicorum*, *G. longipennis*, *G. tabaniformis*, *G. palpalis palpalis*, *G. austeni*, *G. longipalis*, and *G. morsitans orientalis*; and one each of *G. fuscipleuris*, *G. haningtoni*, *G. nigrofusca*, and *G. pallicera*—no blood meal included zebra (Glasgow 1963; see also Jordan, Lee-Jones, and Weitz 1962; Muturi et al. 2011). These findings are mirrored in trypanosome data, with many sites in Africa reporting low infection rates of zebras (e.g., Moloo 1993; Clausen et al. 1998), although occasionally these do occur (Mulla and Rickman 1988). For example, in a study of twenty-four species in the Luangwa valley, Zambia, trypanosome prevalence in plains zebras was 0%, but the 95% confidence intervals were nonzero (0%–12.8%), showing that a few individuals were infected (N. Anderson et al. 2011).

Both sets of studies lend important support to the idea that stripes are instrumental in reducing glossinid feeding. We need this sort of information for other biting diptera—especially tabanids, which are thought to

Table 9.2. Sources of blood food from various hosts in percentages for the three *Glossina* groups.

	Glossina morsitans	Glossina palpalis	Glossina fusca
Domestic			
Human	3.6	13.0	0.5
Cattle	3.6	4.2	2.4
Sheep and goats	0.3	0.4	0.2
Donkey	>0	0.1	>0
Dog	0.2	0.3	>0
Domestic pig	>0	14.0	0.2
Suidae			
Bushpig	5.6	2.2	28.0
Warthog	31.6	1.7	0.4
Bovidae			
Bushbuck	10.2	25.1	11.2
Buffalo	7.5	0.9	7.6
Kudu	5.4		
Others			
Hippopotamus		4.0	15.8
Rhinoceros			14.2
Monitor lizard	>0	6.0	0.1

Source: From Moloo (1993).

be the most important vectors of zoonoses. Tabanids and bloodsucking muscids have painful bites and so are likely to be dislodged quickly and move to another host, potentially spreading diseases rapidly. Moreover, they have large mouthparts that can act as transfer agents (Foil 1989), and blood oozing from bite wounds provides a focal point for other flies that can also carry diseases.

Different dipteran families and species have different landing sites on hosts, as observed in studies of livestock. Generally, lower areas of the host's body are preferred such as the legs and belly (Table 9.3), and this applies also to horses (Muzari et al. 2010). Such regions are especially favored by tabanid species with short labrum lengths, whereas larger tabanids with longer labra prefer upper parts of the host, where the fur is thicker (Mullens and Gerhardt 1979). Ectoparasite-specific site preferences could potentially shape patterns of external coloration across the host's body.

9.3 Ectoparasite population sizes

For biting flies to be a sufficiently strong selective force to drive the evolution of striping, flies must occur in large numbers. Certainly, tabanids, glossinids, simuliids, and mosquitoes do achieve very high densities in some

Table 9.3. Alighting and feeding sites of tabanids expressed as percentages.

Body region	Zimbabwe[a] 16 tabanid spp. 3368 flies	Uganda[b] 20 tabanid spp. 1722 flies	USA[c] 8 tabanid spp. 3419 flies
Legs	61.8	47.7	49.5
Belly and inguinal	15.3	8.2	16.3
Flank	5.8	1.7	10.6
Shoulder	2.1	3.4	3.2
Rump	0.2	9.1	0.7
Back	1.0	5.6	8.2
Neck and dewlap	3.9	9.3	7.1
Head	3.1	1.7	4.3
Tail	0.3	13.1	—
Other	6.4	—	—

[a]Alighting sites of tabanids on cattle in Zimbabwe: Phelps and Holloway (1990).
[b]Feeding sites on cattle in Uganda: Kangwagye (1977).
[c]Feeding sites on cattle in Oklahoma, USA: Hollander and Wright (1980).

areas of the world during certain periods of the year. For instance, mosquito and blackfly outbreaks in certain parts of Canada during the summer are well known. In Africa, local abundances of tsetse flies reached very high numbers prior to control measures (Dorward and Payne 1975): in a review of eight studies reporting a total of eleven estimates, numbers of tsetse flies ranged from 1408 to 105,785 per square mile (Glasgow 1963). On an individual ox, Vanderplank (1944) captured 1455 *Glossina swynnertoni* in Shinyanga Region, Tanganyika Territory, over an eleven-day period in June 1941. Tabanid abundance is far lower but still impressive: maximum figures of 101 *Tabanus taeniola* and 242 *Haematopota vittata* were recorded along a 10,000-yard transect in Bomba, central Tanganyika, in the early 1940s (Glasgow 1946). Densities of other families in Africa are less well documented.

Fly abundance varies both temporally and spatially. Generally, more species of all biting flies are present in the wet than in the dry season, and they occur in greater numbers. Moreover, among East African tabanids there are peaks as to when adults are flying (Vanderplank 1944). Tabanidae are generally found in moist habitats because oviparous tabanids need to breed near mud or damp soil usually close to water, although larvae of different species show great variation in their hydration requirements (Chainey 1993). Even within the same area with similar conditions, tabanid distribution is heterogeneous, as my trapping work revealed.

In contrast, glossinids are viviparous, and larviposition can occur in moist or shady dry areas, so species within each of the three *Glossina* groups are found in many sorts of wooded habitats (Jordan 1993). They are never found in open plains habitats where zebras spend much time grazing, however (Glasgow 1963). Horn flies (*Haematobia*) and stable flies (*Stomoxys*) lay eggs, respectively, on fresh feces, or on moist organic material or vegetation, so are not necessarily found near water (Crosskey 1993b). Mosquitoes lay their eggs in standing water or on damp surfaces (Service 1993). Similarly, biting midges and sandfly larvae require moist, cool places such as leaf litter (Boorman 1993; Lane 1993). In summary, although biting flies as a group are found over many parts of the African continent and throughout much of the three species of zebras' geographic ranges, annoyance will be greater in moister, more humid environments based on ectoparasite breeding biology. It is worth noting that in the comparative analysis described in Chapter 8, striping was associated with six or more consecutive months that fell within a temperature range of 15°C to 30°C and humidity between 30% and 85%, and in the intraspecific analysis stripes was found where temperatures were warm and stable and conditions were wet (Larison, Harrigan, Thomassen, et al. 2015). These situations are conducive to the breeding of many diptera, not just tabanids, and since we do not yet know whether mosquitoes, sandflies, blackflies, or horn flies fail to be attracted to striped objects, we cannot yet dismiss the idea that stripes function to deter several different groups of biting flies.

9.4 Host seeking

For striping to be an antiparasite adaptation, the biting flies must rely on vision to find their host. Most of these dipteran groups are anautogenous, meaning that the female requires a blood meal for maturation of her ovaries, whereas males do not need blood for reproduction. Glossinids and bloodsucking muscids are exceptions, in that both sexes require blood meals to reproduce. Female biting flies are first thought to locate prey by flying up an odor plume either within it or along its periphery and to employ vision only when they are close to their target, but this is highly variable across families (Lehane 2005). Carbon dioxide, acetone, and octenol can all stimulate insect activity at a distance and act as attractants to mosquitoes (Gillies 1980), midges (Bhasin, Mordue [Luntz], and Mordue 2000), blackflies (Sutcliffe 1986), and tabanids (Zollner et al. 2004), whereas phenylpropanoids attract *Stomoxys calcitrans* (Hammack and Hesler 1996). Odors may be detected 90 m away by some families, but generalizations are risky because odors have different effects on each species (Sutcliffe 1987; Lehane 2005). For example, horse urine increases capture of tabanids *Tabanus bromius* 1.6-fold but *Alylotus quadrifarius* 3.5-fold living in the same habitat in southeastern France (Baldacchino et al. 2013).

Once a dipteran is close to a host, it may switch to visual cues (Allan, Day, and Edman 1987; Sutcliffe 1987; Bidlingmayer 1994), but it is worth remembering that flies have poor resolution compared to humans, seeing objects from only a few meters away (see Bidlingmayer and Hem 1980; but see Napier Bax 1937). Object detection depends on differences in color, brightness, and shape in relation to the background (Allan, Day, and Edman 1987). Ultraviolet and blue-green parts of the spectrum are most attractive, as too is black (Lehane 2005; Santer 2014). Furthermore, objects that have maximum intensity contrasts compared to the background, that are three-dimensional, that have nonlinear shapes, and that show movement are all attractive (Lehane 2005). Yet the relative import of these factors varies by species and whether diptera seek hosts at night or during the day; indeed, one reason that stripes are thought to act against glossinids and tabanids is that these groups are highly diurnal (G. Gibson and Torr 1999). The ways in which horseflies, stable flies, and tsetse flies fail to recognize striped objects is still very speculative. Ideas include stripes selectively obliterating the body's edge, reducing overall contrast with the background, increasing the number of edges so as to disguise the object's outline, or diminishing the degree or direction of linear polarization (section 5.8.b) or disturbing optomotor responses for landing; some of these ideas are difficult to reconcile with glossinids' (and blackflies') high sensitivity to contrast (Turner and Invest 1973; Browne and Bennett 1980). I see investigation of biting dipterans' optomotor responses to stripes as being a critical next step in understanding why zebras are striped.

9.5 Parasites and diseases transmitted by bloodsucking diptera

One of the reasons to avoid attack by biting flies is to reduce the probability of contracting a dangerous or fatal disease, and biting flies transmit a large number. Infamous ones include lieshmania protozoa by sandflies; plasmodium malaria, filarial worms, yellow fever, dengue fever, and equine encephalomyelitis by mosquitoes; onchocercias caused by a filarial nematode, *Onchocerca volvulus*, carried by blackflies; several viral diseases, including African horse sickness, by biting midges; equine infectious anemia virus and anthrax by tabanids; and trypanosomiasis by glossinids. As experimental studies have explicitly described tabanids', glossinids', and *Stomoxys'* difficulties in landing on striped surfaces, it is worth documenting the viruses, bacteria, protozoa, and helminthes carried by these groups (see Table 9.4). A number of these ailments are very trouble-

Table 9.4. Disease agents associated with tabanids and *Stomoxys*.

Tabanids	*Stomoxys*
Viruses	
Equine infectious anemia	Equine infectious anemia
Vesicular stomatitis	Yellow fever
Hog cholera	African horse sickness
Rinderpest	Poliomyelitis
California encephalitis	
Western equine encephalitis	
Tick-borne encephalitis	
African horse sickness	
Foot and mouth disease	
Rabbit myxoma	
Bacteria	
Anaplasma marginale	*Borrelia* sp
Coxiella burnetii	*Leptospirosis*
Bacillus anthracis	*Bacillus anthracis*
Clostridium chauvoei	*Brucella* sp
Pasteurella multocida	*Erysipelothrix* sp
Francisella tilarensis	*Salmonella typhimurium*
Brucella sp	Streptotricosis
Listeria monocytogenes	
Erysipelothrix rhusiopathiae	
Leptospira sp	
Clostrdium perfringens	
Fusobacterium necrophorum	
Potomac horse fever	
Protozoa	
Babesia ovate	*Trypanosoma evansi*
Besnoitia besnoiti	Toxoplasmosis
Haemoproteus metchnikovi	Leishmaniasis
Trypanosoma theileri	
Trypanosoma evansi	
Trypanosoma equiperdum	
Trypanosoma vivax[a]	*Trypanosoma vivax*[a]
Trypanosoma congolense[a]	*Trypanosoma congolense*[a]
Trypanosoma brucei brucei[a]	*Trypanosoma brucei*[a]
Trypanosoma brucei gambiense[a]	
Trypanosoma simiae[a]	
Theileria tarandi rangiferis[a]	
Helminths	
Loa loa	*Setaria labiatopapillosa*
Dirofilaria repens	*Stephanofilaria stilesi*
Dirofilaria roemeri	Habronemiasis
Elaeophora schneideri	
Setaria equina	
Nematoda	
Onchocerca gibsoni	

Sources: Disease agents associated with tabanids from Krinsky (1976); Foil (1989). Disease agents associated with *Stomoxys* from Zumpt (1973).

[a]Denotes glossinid-borne diseases.

Table 9.5. Trypanosome infection rates in *Glossina swynnertoni* (by microscopic examination) within and around Serengeti National Park, Tanzania.

Trapping sites	Total dissected	Percentage infected	Trypanosome types		
			vivax	*congolense*	*brucei*
Retima hippo pool	116	0	0	0	0
Sopa	42	9.3	2	2	0
Kilimafedha	139	13.7	16	0	3
Simuyu post	19	5.3	1	0	0
Death Valley	98	7.1	1	2	4
Robanda/Ikoma Gate	122	7.4	1	2	6
Makao	141	16.3	8	8	7

Source: From Malele et al. (2007).

some to African equids, and four stand out as being lethal: trypanosomiasis, equine infectious anemia, African horse sickness, and equine influenza (Table 8.3). Zebras are highly susceptible to anthrax too as judged by absence of seropositivity and frequent fatalities, at least in the Serengeti (Lembo et al. 2011), but we do not know which of these (or other diseases) are the key sources of mortality for zebras in the wild. Moreover, it is possible that other ectoparasite groups are potent vectors of these and other diseases harmful to wild equids, and it is possible too that these diptera fail to recognize striped surfaces as hosts.

The incidence of diseases in biting fly populations may have a strong bearing on the extent of striping in zebra populations, but quantitative data on numbers of individual flies that carry diseases fatal to zebras are sparse and are chiefly limited to trypanosomiasis in tsetse flies (see Table 9.5 for an illustration). Certainly, temperature may be an important factor affecting infection rates in both dipterans and mammalian hosts (e.g., Welburn and Maudlin 1991). If striping deters biting flies' landing responses, then we should expect low incidences of diseases in zebras compared to sympatric herbivores. Reports suggest low incidence of trypanosomiasis in some areas (Table 9.6), the disease for which most surveys of mammalian hosts have been carried out.

9.6 Mechanistic studies

These aspects of vector biology open up a series of important mechanistic questions, studies of which would solidify the hypothesis that ectoparasites are the evolutionary driver of zebras' external pelage patterns. I con-

Table 9.6. Some surveys for trypanosome infections in wild animals.

Location	Overall trypanosome prevalence (%)	General remarks	Source
Luangwa Valley, Zambia	28.0	16% with *T.b. rhodesiense*. Bushbuck and waterbuck most commonly infected.	[a]
Luangwa Valley, Zambia	23.0 (43 shot)	Bushbuck and waterbuck most commonly infected.	[b]
Luangwa Valley, Zambia	14.5 (including 9 mixed infections)	*T.b. rhodesiense* detected in waterbuck and warthog.	[c]
Luangwa Valley, Zambia		*T.b. rhodesiense* detected in zebra and impala.	[d]
Botswana	15.7 (in buffalo, lechwe, and reedbuck)	Highest prevalence in buffalo occurred at 2.5 years and then dropped.	[e]
Serengeti, Tanzania	7.5 (infection in mammals)		[f]
Comoé National Park, Côte d'Ivoire		Highest trypanosome prevalence in *Hippotragus*, buffalo, and waterbuck.	[g]
Busoga, Uganda		Trypanosome infections in bushbuck, waterbuck, and elephant.	[h]
Queen Elizabeth National Park, Uganda	11.0	Highest prevalence in bushbuck.	[i]
Lambwe Valley, Kenya	16.0	90% prevalence in bushbuck.	[j]
Guvoro and Mutuali, Mozambique		3% in bushbuck, buffalo, and wildebeest; 2% in southern reedbuck, greater kudu, oribi, and waterbuck.	[k]
Ukerewe, Shinyanga, Tanzania		Greater kudu, Coke's hartebeest, and giraffe but sample sizes low. One of four zebra positive.	[l]

[a]Kinghorn and Yorke (1912a); Kinghorn and Yorke (1912b).
[b]Keymer (1969).
[c]Dillmann and Townsend (1979).
[d]Mulla and Rickman (1988).
[e]Drager and Mehlitz (1978).
[f]Bertram (1973).
[g]Komoin-Oka et al. (1994).
[h]Burridge et al. (1970).
[i]Mwambu and Woodford (1972).
[j]Allsopp (1972).
[k]de Andrade and da Silva (1959).
[l]Vanderplank (1942).

sider answers to these questions as further keys to advancing understanding of the evolution of stripes in the three species of zebras (Table 9.7). First, we need to establish whether experimental findings with tabanids and glossinids and *Stomoxys* using striped targets, traps, and models with and without odor can be replicated with real animals under field conditions and whether these extend to other families of biting flies. We also need to confirm whether it is striping itself, a two-tone coat, or even extent of white pelage that protects equids from biting flies. Because of the difficulties in counting and identifying flies at distance, it may be necessary to capture flies landing on tame zebras, and compare them to those land-

Table 9.7. Some remaining questions.

Host choice
Which host species are most attractive to biting flies?
Numbers
What are absolute numbers of ectoparasites over time and space?
Host seeking
How do stripes thwart biting fly attack?
Infectious diseases and blood loss
What is the incidence of diseases in biting fly populations and zebra populations?
What are the effects of blood loss on zebras?

ing on white horses, black horses, and brown control animals; horses are useful because they are more easily manipulated than other nonstriped equids. Alternatively, tame zebras might be dyed black or white, but this may be difficult logistically and there may be structural differences between white and black hairs that could still influence reflectance. Ideally, any such study needs to be conducted in Africa, where ectoparasites of striped equids are found. Certainly, there are horse farms in southern Africa with tame zebras, but difficulties lie in sampling a diversity of biting fly species and genera across the continental range of zebras. Other alternatives include closeup photography of live animals in the wild recording numbers of flies from photographs (see Horvath et al. 2010) or sampling blood meals of diptera themselves, but these latter field studies do not use nonstriped equid controls (see Clausen et al. 1998).

Second, we need to know the numbers of flies landing on individual zebras over time and space in order to verify whether ectoparasites can impose severe selection pressure on zebras. I am concerned that despite the association between striping and months of favorable conditions for tabanids, and between belly stripe number and glossinid distributions (section 8.2.d), plains zebras actually spend considerable portions of the year feeding on dry, grassy plains if they are available. This is an unfavorable habitat for tabanids that prefer moist closed conditions (Foil and Hogsette 1994) and for tsetse flies that rest on trees. Zebras avoid floodplains in the wet season and move to woodlands where both fly families are plentiful especially in that season, but it would be helpful to know the extent to which zebras select habitats to avoid attention of biting flies. Such a study will involve sampling families of flies across wet and dry seasons and across different habitats; it is not trivial because each dipteran family is captured in different sorts of traps, necessitating considerable trap effort.

Third, we need to understand why certain diptera appear unable to land on striped objects. Biting flies probably become aware of the approximate location of a host using olfactory information; they move fast, at less than a meter above ground; and only employ visual cues very close to the target. Indeed, we know that some biting diptera fly past their host before circling back (Allan and Stoffolano 1986a). Somehow stripes viewed close up prevent horseflies, tsetse flies, and stable flies and perhaps other groups from landing on the host. Do the stripes cut up the target into thin strips, preventing object recognition? Is the outline of the zebra disrupted because black stripes blend in with the dark background? Or do white stripes reduce the luminance contrast with the background? It would be helpful to predict the distances at which stripes of differing widths can be resolved based on ommatidial dimensions and distributions in the eyes of different biting fly species, as attempted by Turner and Invest (1973) and Britten and colleagues (2016), and to supplement these studies with electrophysiological recordings and anatomical studies (e.g., W. Smith and Butler 1991).

Fourth, we are not yet clear about the role of linearly polarized light in attracting biting flies to zebra pelage or deterring them from it. Horvath's group argues that white hairs scatter light, affording no polarization signature, whereas black hairs reflect horizontally polarized light strongly, especially when the sun is low (Horvath et al. 2010). On the other hand, we have shown that white and black stripes have a polarization signature and that differences between them are annulled at 10 m away (Britten, Thatcher, and Caro 2016). I am personally nervous of assuming a strict dichotomy to stripes' polarization signatures when, at any one point in time, only some areas of the body will be oriented to reflect horizontally polarized light; some of the hairs within white stripes have brown phaeomelanin pigment; white stripes are actually cream-colored or yellowy in some populations; white stripes are often dirty due to rolling in dust or standing in mud (Plate 9.1); and the coat may be wet during rain or covered in fine dust during the dry season. Moreover, we do not yet have the behavioral evidence to know which families of biting diptera are polarotaxic other than tabanids. Anatomical and limited electrophysiological evidence suggest R7 and R8 cells in tsetse flies' retinas can detect polarized light (Hardie, Vogt, and Rudolph 1989), and the marginal ommatidia in *Calliphora* and *Musca* (flies that can cause considerable discomfort to domestic horses) are sensitive to the E-vector of polarized light (Wunderer and Smola 1982; Hardie 1984; Hardie 1986; Hardie, Franceschini, and McIntyre 1979). Possibly, different aspects of reflected light ward off different groups of ectoparasites, or defeat even a single species of biting fly. If so, what is the relative importance of these mechanisms?

Fifth, we are not sure why biting flies are such a problem for equids. There are at least three possibilities: they carry dangerous equid diseases, blood loss, or simply mechanical annoyance that interferes with foraging (Horvath et al. 2010). Currently, diseases seem the most likely candidate explanation in that tabanids are vectors for at least four diseases fatal to zebras (Table 8.3), and these are restricted to Africa, where striping is seen. Critically, we need to understand the extent to which tabanid and glossinid populations carry these diseases and how these vary geographically across zebras' ranges (Njiru et al. 2004; Larison, Harrigan, Thomassen, et al. 2015). For example, tsetse fly sex, age (and therefore fly age structure), fly hosts, and temperature are all known to influence tsetse infection rates with trypanosomes (Jordan 1976; Welburn and Maudlin 1999). Under severe hot and dry conditions, many flies do not live long enough to develop mature infections of trypanosomes. Unfortunately, no comprehensive characterization of trypanosomes has been conducted in wild animals, yet surveys of the incidence of these diseases in free-living zebra populations are required to understand how many individuals test positive for infection and how many survive. Lethal surveys are impossible in protected areas, or for Grevy's and mountain zebras, which are Endangered and Vulnerable, respectively: nonintrusive methods are needed such as passive serosurveillance (see Lembo et al. 2011).

We can calculate absolute blood loss from studies of livestock by measuring before/after blood meal weights of flies (which incidentally vary considerably among diptera [Wiesenhutter 1975]) and multiplying up by numbers of flies. Measurements need to be conducted in different seasons and habitats. Effects of blood losses on milk production (e.g., Grannett and Hansens 1956; Taylor, Moon, and Mark 2012) and work carried out by horses (e.g., Riha et al. 1983) have been calculated and can perhaps be extrapolated to zebras.

Finally, we need to assess the extent to which mechanical disturbance from bites of tabanids and bloodsucking muscids reduces the time that hosts spend feeding. The cutting mouthparts of these flies pool blood in the epidermis and cause considerable pain to humans. We know that domestic horses vary in their ability to withstand biting fly annoyance, with older individuals being less disturbed (Foil 1989), but this needs to be quantified in zebras too. Again, observations of tame zebras and differently colored horses feeding would help us understand the extent to which biting flies and other diptera reduce food intake.

9.7 Further outstanding issues

9.7.a Multiple functions

When there are a number of competing hypotheses for the function of a trait, there is a temptation to ask whether the trait has more than one function in the sense of the character having been molded and currently being shaped by more than one selection pressure. External coloration is likely to be a compromise of sorts because feathers, scales, hairs, and exoskeletons of different colors necessarily absorb different wavelengths of light and therefore have ramifications for heat load, but these may not be of primary importance. Larison, Harrigan, Thomassen, and colleagues (2015) found an association between striping in plains zebras and constant temperatures, warmer temperatures in the cold season, and high rainfall, but these might be a proxy for fly infestation or disease infection rates. Instead of suggesting that there may be both temperature and ectoparasite benefits, I believe it is most parsimonious to adhere to the ectoparasite hypothesis because of experimental evidence that tabanids eschew landing on striped surfaces (e.g., Egri, Blaho, Kriska, et al. 2012; Blaho et al. 2013); that stripe widths are thinner than those on which horseflies, stable flies, and tsetse flies are prepared to land; that zebra pelage is thinner than the length of biting fly mouthparts, making zebras susceptible to blood loss; and because biting flies carry diseases fatal to zebras (Caro et al. 2014). In contrast, there is no evidence as yet for or against striping setting up convection currents over the dorsa of zebras or when this might occur.

Leaving thermoregulation aside for the moment, we can ask whether stripes have evolved not only to thwart biting fly attack but as an antipredator adaptation or for intraspecific signaling. My view is that there is no other primary function, because I view intraspecific signaling and predation hypotheses as untenable (Table 9.1). Yet there are two passing worries: spotted hyenas, an abundant predator across the African continent, take zebras less than expected from zebra abundance (Hayward 2006) and so might possibly be thwarted by stripes. Second, there are significant associations between extent of striping and the geographic range of spotted hyenas in regard to rump striping and leg stripe intensity at the species level, although associations did not achieve significance at the subspecific level, which is a more sensitive analysis (section 8.2). Possibly, then, rump and leg striping confuse or act as an aposematic signal to spotted hyenas, but I am loath to entertain these possibilities given relative size differences of predator and prey: a spotted hyena weighs just 16.7%–29.6% of an adult plains zebra, 15.1%–26.8% of an adult mountain zebra, and 11.3%–20.0% of an adult Grevy's zebra (call it a sixth to a quarter as much). Across carnivores, species more than 21.5 kg in body weight (like a spotted hyena) feed mostly on prey that is greater than 45% of their own mass (i.e., about half their size) (Carbone et al. 1999), although large carnivores in Kruger National Park favor prey one half to twice their mass (Owen-Smith and Mills 2008). Even if we double these relative weights that can be toppled to account for spotted hyenas hunting in groups (i.e., the same to even four times the mass of a hyena), we are still left with the fact that adult zebras are above or on the cusp of prey sizes that spotted hyenas can manage. Thus, large body size of zebras provides the most parsimonious explanation for "avoidance" of this prey item rather than stripes. Two further considerations are noteworthy: species-level analyses of striping were conducted on just a small sample size of only seven species, and striping evolved only once in the clade. Both facts make our comparative analysis at the species level a less than compelling reason for accepting that striping is being driven by hyena predation as well as by biting flies.

9.7.b Loose ends

Additionally there are a series of unanswered questions that nag those interested in the evolution of mammalian coloration. First, why are zebras not completely white? White pelage attracts far fewer flies than black pelage in domestic horses (Horvath et al. 2010), and research shows that stripes are only marginally more effective at deterring flies than white surfaces (Egri, Blaho, Kriska, et al. 2012). Perhaps all-white pelage carries too great a cost on the open savannahs of Africa? (Actually, very few herbivores are pale or creamy-white outside of desert environments [Stoner, Caro, and Graham 2003]). Costs of white coats might include poor background matching or an inability to absorb radiation on cool days. Alternatively, absence of melanin could compromise photoprotection against harmful ultraviolet radiation or fail to protect against mechanical abrasion (Zeise, Chedekel, and Fitzpatrick 1995). If large patches of black hair attract biting flies, perhaps two-tone pelage is a compromise solution to reduce fly attack but avoid costs of white pelage?

This directly raises a second question of why zebras need to be striped rather than sport patches of black and white pelage. Blaho and coworkers (2012) have shown that black spots on a white background afford protection from tabanids with greater effects as spot size decreases and spot number increases. This is an important finding. They interpreted these findings as spotty surfaces reflecting less linearly polarized light and at different angles than a black surface. But being spotted will also reduce overall luminance compared to the background. Whatever the mechanism, Blaho and colleagues' experiments suggest that stripes per se are not the key to stopping tabanids landing on an object but that breaking up the black surface is important. So what is special about stripes? Are equids genetically predisposed to develop stripes rather than spots? Stripes on the legs of some African wild asses and Przewalski's horses suggest this might be the case. Certainly, comparisons of alighting preferences on spotty and striped targets would be very useful and could lead to a better understanding of how dipteran vision functions.

Third, why do some subspecies of mountain and plains zebras and Grevy's zebra have white or light ventra? The position on the body makes it unlikely these are required to reflect radiation and cool the animal. Perhaps instead, if they are not involved in self-shadow concealment, as our preliminary data indicate (section 2.3), bellies may be white simply to circumvent costs of producing of black stripes? In other words, that there are physiological costs of eumelanin production in black pelage. Perhaps melanin sequestered in hair is not available to defend cells from oxidative stress, which compromises immunocompetence, as noted in some birds (Griffith, Parker, and Olson 2006); similar data on mammals are very scarce indeed. Most discussions of the functions of striping in zebras to date focus on possible benefits but avoid considering developmental, maintenance, or injury costs.

Fourth, why do Grevy's zebras have thin stripes whereas mountain and plains zebras have far thicker stripes on their necks, flanks, and rumps? Is this because Grevy's zebras are encountering differing genera of biting flies with different eye anatomies, or are the reasons nonadaptive; in other words, would any striping pattern suffice? Nonadaptive (nonfunctional) explanations are unsatisfactory and difficult to demonstrate on account of proving a negative, but they need consideration here: recall that the African wild ass has jagged black thin stripes on its lower legs set against a gray background, which seem sufficient to ward off tabanids cruising half a meter above ground (Allan and Stoffolano 1986a).

Fifth, why do zebra foals have brown tints to their coats (Plate 9.2)? These remain for only a matter of weeks, but the blurring of dorsal stripes and the ruddy color seem designed as background matching against brown grasses (although zebras are born year round). There are reports that tabanids land far less on horse foals than on adults (Foil et al. 1985), related perhaps to reduced carbon dioxide or odor production. If this is the same for zebras, and there are no data as yet, it might explain why zebra foals can afford to wear cryptic coats if this is indeed the reason for light brown pelage. This topic needs investigation.

Last, given that artiodactyls suffer from some of the same diseases as equids and are annoyed by biting flies too, why are they not striped? We have argued that this is because zebras have thinner pelage than heterospecifics (section 8.2.d), but we cannot separate cause and effect. Do striped coats allow their bearers to have thinner pelage, or does thinner livery select for striped coats to keep flies away? If the latter, then why do zebras have thin coats? One artiodactyl is striped. The okapi has similar stripes on its rump and upper legs as does a plains zebra but has white lower legs (Plate 9.3). What are the selection pressures driving this pattern of external coloration (the obverse of that seen in the quagga) in the forests of eastern Democratic Republic of Congo? We really have no idea.

9.8 Conclusion

Sex allocation theory, which is concerned with the ways in which sons and daughters are produced so as to maximize parental reproductive success, was slow to be accepted by the scientific community because the mechanisms by which mothers bias the sex ratios of their progeny were opaque (Silk and Brown 2008; S. West 2009). The same problem applies to biting flies being the evolutionary driver of zebra stripes. We now need to understand mechanisms of host choice and attraction across a broader range of dipteran taxa, quantify biting fly numbers, understand biting fly vision better, rates of vector infection, the diseases involved, and their effects on zebras and other African herbivores. In addition, there are a number of fine-grained questions about aspects of striping that remain outstanding. There is still a lot of work to do: let's get to it.

Figure bm.1. Occasionally, zebras are born with strange pelage that can give clues to the development and function of coat coloration. That so few aberrant individuals are found in the wild suggests that striping is under strong selection. (Drawing by Sheila Girling.)

Appendix 1

Scientific names of vertebrates mentioned in the text.

PISCES

Order Orectolobiformes
Zebra shark   *Stegostoma fasciatum*
Order Anguilliformes
Zebra moray   *Gymnomuraena zebra*
Order Clupeiformes
California anchovy   *Engraulis mordax*
Order Gasterosteiformes
Three-spined stickleback   *Gasterosteus aculeatus*
Order Perciformes
Zebra red dorsal   *Metriaclima pyrsonotos*

REPTILIA

Order Squamata
Desert banded snake   *Simoselaps bertholdi*
California king snake   *Lampropeltis getula*

AVES

Order Falconiformes
Fish eagle   *Haliaeetus vocifer*

MAMMALIA

Order Primates
Indri   *Indri indri*
Angolan black and white colobus   *Colobus angolensis*
Order Carnivora
Gray wolf   *Canis lupus*
African wild dog   *Lycaon pictus*
Fennec   *Vulpes zerda*
Cheetah   *Acinonyx jubatus*
Lion   *Panthera leo*
Leopard   *Panthera pardus*

Tiger   *Panthera tigris*
Spotted hyena   *Crocuta crocuta*
Brown hyena   *Hyena brunnea*
Order Proboscidia
African elephant   *Loxodonta africana*
Order Perissodactyla
Domestic horse   *Equus caballus*
Przewalski's horse   *Equus ferus przewalskii*
Plains zebra   *Equus burchelli*
Mountain zebra   *Equus zebra*
Grevy's zebra   *Equus grevyi*
African wild ass   *Equus africanus*
Donkey or ass   *Equus asinus*
Kiang   *Equus kiang*
Asiatic wild ass   *Equus hemionus*
Malayan tapir   *Tapirus indicus*
Black rhinoceros   *Diceros bicornis*
Order Artiodactyla
Warthog   *Phacochoerus aethiopicus*
Bushpig   *Potamochoerus larvatus*
Hippopotamus   *Hippopotamus amphibious*
Giraffe   *Giraffa camelopardalis*
Okapi   *Okapia johnstoni*
Impala   *Aepyceros melampus*
Hartebeest   *Alcelaphus buselaphus*
Blue wildebeest   *Connochaetes taurinus*
Topi   *Damaliscus korrigum*
Thomson's gazelle   *Gazella thomsonii*
Oribi   *Ourebia ourebi*
Steenbok   *Raphicerus campestris*
Sharpe's grysbok   *Raphicerus sharpei*
Cape buffalo   *Syncerus caffer*
Eland   *Taurotragus oryx*
Bongo   *Tragelaphus eurycerus*
Bushbuck   *Tragelaphus scriptus*
Greater kudu   *Tragelaphus strepsiceros*
Zebra duiker   *Cephalophus zebra*
Bush duiker   *Sylvicapra grimmia*
Roan antelope   *Hippotragus equinus*
Sable antelope   *Hippotragus niger*
Beisa oryx   *Oryx gazella*
Waterbuck   *Kobus ellipsiprymnus*
Puku   *Kobus vardoni*
Reedbuck   *Redunca redunca*

Appendix 2

Nature of wounding seen in African ungulates in Katavi National Park. Some individuals received more than one wound.

ZEBRA
2009
Tail
No tail; cut off at base
Tail cut in half
Half tail remaining
Rump
4" scar below rump
6" scar high up near tail
8" old line scar at bottom of rump on side
2" old nick + 1" nick on top of rump
7" × ½" long rump scar + 2 thin long lines (1 is 10" long)
2" circular patch on midrump
Vertical 4" scar on lower rump
Odd scratches on rump
2" vertical old scar on midrump + one on stomach low down near rump
9" vertical gash from midrump to top of leg
6" horizontal cut on rump
18" vertical gash down hindleg
18" diagonal gash across left rump + white circle at top
24" vertical cut on rump down to top of leg
Flank
Four lacerations on flank
Parallel scratches on stomach
3" horizontal old scar on stomach

Other

Old slash between shoulder and top of foreleg

Lower mane missing + multiple old cuts on neck and flank

Oval-shaped 3" × 2" scar

12" scar

6" long vertical old scar

2010

Tail

Broken tail 2" from base + point scar on right haunch

Half hair of tail missing

Half tail missing

Half tail missing

Rump

Large circular healed gash from top of tail down left haunch + 4" × 1½" fresh gash on lower right haunch

7" curved scar midway down rump

Black healed 8" long scar at top of haunch 2" wide at front, 1" wide at rear + half tail

Old 10" diagonal slash on haunch

Three 3" old cuts on flank and haunch

6" healed vertical scar on left rump

6" healed scar on upper left haunch

Three 2" scratches on rump

2" graze on flank + 2" graze on haunch

9" thin line between hindleg and flank (not sure if scratch or mud)

15" healed diagonal scar on side of hindleg

8" healed scar diagonally on top of hindleg

12" long diagonal scar on haunch

2" vertical scar on haunch

Jagged 18" healed scar on top of haunch

18" oblique healed scar on top of haunch

24" scar + ¼ tail missing + fresh wound on other haunch

7" old oblique scar on haunch

Flank

Small cross on flank (might not be a wound)

2011

Tail

6" tail

Tail hairs missing

8" tail

Rump

12" horizontal gash along left haunch + 4" × 4" circular scar at end + 6" tail

2" vertical old scar on rump

Two black patches on top of rump

24" old diagonal line on left haunch

21" old scar across haunch

Blemish on right haunch

7" old diagonal line on lower haunch

4" old scar on front part of haunch

6" old closed gash on central part of haunch

8" upwardly curving old scar on central part of haunch

Old white cut through a black stripe on haunch

Zigzag black stripe that may have had a scar on it on haunch

4" diagonal old scar halfway down rear of haunch

4" diagonal gash on lower haunch—fresh at distal end

6" × 2" old scar on front of haunch

Flank

8" vertical scar from just behind groin + 2" vertical line on flank

Other

Limping left foreleg

GIRAFFE

2009

Tail

Tail cut in half

12" tail only

Straggly hairs on tail (not sure if wound)

Rump

Old ½" horizontal nick at tail base

7" curved scar on low thigh

4" vertical old cut at front of hindleg

2" old scar in front of rump + kink in tail

Old thin line on top of hindleg

3" healed light gash on midhindleg

Flank

3" horizontal scar on lower flank

Slight nick on stomach

Horizontal line on stomach

Two thin 6" scratches near top of back (not likely a predator)

Rear of stomach a wide, dry sore

6" slight gash on midflank

Other

2" circular nick in front of hindleg

4" × 3" healed scar with raw skin on front of top of right foreleg

6" × 4" area with flesh removed

2010

Tail

Tail missing all hairs and maybe some flesh—old

Tail only 12" long

18" tail

Rump

3" old cut on hindleg 6" above "knee" + 3" × 1½" scar at top of foreleg

Scratch (or mud) on right haunch

9" healed vertical gash on haunch

Irregularly shaped 4" × 3" lump on haunch (not sure if scar or growth)

2011

Other

2" × 2" old scar front foreleg + old 2" × 1" scar on lower neck + 10" × 4" diamond-shaped scar with outside red but inside black scar tissue on left shoulder

IMPALA

2009

Rump

Limping hindleg

Flank

3" black line along midflank (not sure if scar or dirt)

2010

Flank

White spot on center of flank (not sure if scar)

Scratch (or dirt) on flank

2011

Rump

3" line on haunch with spots (not sure if scar)

Flank

Diamond-shaped 7" × 3" piece of wood embedded in stomach

5" line at border of dark dorsum/light ventrum behind foreleg

3" white fur parallel to top of dorsum on flank area

Other

Right foreleg broken

Broken right horn halfway down

BUFFALO

2009

Tail

4" tail

3" tail

No tail

Rump

12" × 4" old scar on left rump

Flank

　　Large white circular hole two-thirds along the back

　　Old gash low on flank

Other

　　Left horn broken halfway down

2010

　Tail

　　　3" tail

　　　8" tail

2011

　Tail

　　　6" tail

　Other

　　Missing half of left horn + three white patches on front of flank

WARTHOG

2010

　Tail

　　　2" tail

Appendix 3

Families of insects identified in each type of biconical trap color.

		Biconical trap colors			
	Blue	Dark brown	Light brown / black	Light brown	Striped
Acrididae	1				
Aphididae		1			
Bethylidae		1	1		
Calliphoridae		2	1	3	2
Carabidae					1
Chloropidae			1		
Chrysomelidae	1	2			
Cixiidae			1		
Dolichopodidae	1		1		
Drosophilidae		1			
Elateridae		1			
Empididae			1		
Formicidae	1				
Glossinidae	2	1	1	1	1
Ichneumonidae	1				
Lepidoptera	1				
Lygaeidae	1				
Miridae					1
Muscidae	5	4	6	6	4
Mycetophilidae		1		1	
Nitidulidae	1				
Sarcophagidae				2	
Stratiomyidae	1		1		1

	Biconical trap colors				
	Blue	Dark brown	Light brown / black	Light brown	Striped
Syrphidae	1	1	1	2	
Tabanidae	1	1		2	
Tachinidae				1	
Tephritidae	1	2	3	3	5
Tettigoniidae	1				1
Ulidiidae			1		1
Totals	15	12	12	8	10

Source: Danielle Whisson, unpublished data.

Note: Numbers refer to more than one morphotype. These are minimum estimates, as individuals misidentified as having already been taken as a specimen may not have been collected.

Appendix 4

Families of insects identified in each type of cloth trap color.

	Cloth trap colors				
	Uniform black	Uniform gray	Uniform white	Horizontal striped	Vertical striped
Agromyzidae	X	X		X	X
Anobiidae	X		X		
Anthomyiidae				X	
Anthomyzidae				X	
Aphididae	X				
Bethylidae	X		X	X	X
Berothidae			X		
Bombyliidae			X		
Blattodea				X	
Braconidae	X	X	X	X	X
Buprestidae			X		
Calliphoridae	X	X	X	X	
Cantharidae	X				
Carnidae		X			X
Cercopidae			X	X	X
Chalcididae	X	X	X	X	X
Chironomidae	X				X
Chloropidae		X			X
Chrysomelidae	X	X	X	X	X
Chrysididae	X			X	
Cicadellidae	X	X	X	X	X
Cixiidae	X		X	X	
Coccinellidae	X	X	X	X	

	Cloth trap colors				
	Uniform black	Uniform gray	Uniform white	Horizontal striped	Vertical striped
Colydiidae				X	
Crabronidae	X			X	X
Curculionidae	X	X	X	X	X
Delphacidae		X			
Diptera*	X	X	X	X	
Dolichopodidae	X	X	X	X	X
Drosophilidae	X			X	X
Elateridae	X		X	X	
Empididae	X		X		
Encyrtidae			X	X	X
Eucharitidae			X		
Eupelmidae				X	X
Eurytomidae				X	
Evaniidae	X	X		X	X
Formicidae	X	X	X	X	X
Glossinidae	X	X	X		X
Ichneumonidae	X	X			X
Lepidoptera	X	X	X	X	X
Lygaeidae				X	X
Mantidae			X		
Melyridae			X	X	
Milichiidae					X
Miridae	X	X	X	X	X
Mordellidae	X				
Muscidae	X	X	X	X	X
Mycetophagidae					X
Mycetophilidae	X	X	X	X	
Nitidulidae				X	X
Phalacridae		X	X	X	X
Platypodidae				X	
Pompilidae	X	X			
Psyllidae	X	X	X	X	X
Pteromalidae	X	X	X	X	
Salticidae	X				
Sarcophagidae	X				
Scarabidae				X	
Sciaridae	X				
Scelionidae				X	
Sclerogibbidae	X				

	Cloth trap colors				
	Uniform black	Uniform gray	Uniform white	Horizontal striped	Vertical striped
Staphylinidae	X				
Stratiomyidae	X	X	X	X	X
Syrphidae					X
Tabanidae	X	X	X		X
Tachinidae	X	X	X	X	X
Tenebrionidae	X	X	X	X	X
Tenthredinidae					X
Tephritidae	X	X	X	X	X
Thrips†	X				
Tipulidae		X	X		
Xylophagidae		X			X
Totals	44	32	36	43	37

Source: Danielle Whisson, unpublished data.

Note: Several morphotypes can contribute to one family. These are minimum estimates, as individuals misidentified as having already been taken as a specimen may not have been collected.

*Diptera—possible Muscidae but specimens poorly preserved.

†Order Thysanoptera but could not be identified to family.

Appendix 5

Photographic sources for comparative analyses.

Photographs used for scoring equid pelage were obtained from reputable sources (see the list below) on the Internet and in print. "Reputable" sources are those such as encyclopedic websites, published mammalogy books, zoo websites, national park websites, and other peer-reviewed sources; no image was taken from an individual's personal or unidentified website. Only photographs with accurately identified animals where the entire animal could be seen were used. Only one animal per photograph was scored. Sources used are:

Plains Zebra: Damara (*Equus burchelli antiquorum*)
Macdonald (1984), p. 120
Groves (1974), p. 72
Madamzebra.com
flickr.com
123rf.com
biology4kids.com
oregonzoo.com
zooinstitutes.com (Cincinnati)
zooinstitutes.com (Izhevsk)
zooinstitutes.com (Riga)
zooinstitutes.com (Philly)
zooinstitutes.com (Kerkrade)

Plains Zebra: Burchell's (*Equus burchelli burchellii*)
Macdonald (1984), p. 109
theequinest.com
petermass.nl (photo of specimen from Berlin's Museum fur
    Naturkunde)

biopix.com
treknature.com
pawsforwildlife.co.uk
biolib.cz
calphotos.berkeley.edu
wikipedia.org
commons.wikimedia.org

Plains Zebra: Chapman's (*Equus burchelli chapmani*)
Macdonald (1984), p. 132
shoarns.com
zoobarcelona.cat
theequinest.com
wikipedia
zoochat.com
de.wikipedia.org
zoo-safari.com.pl (upper)
jameshagerphoto.com
biolib.cz
arkive.org

Plains Zebra: Crawshay's (*Equus burchelli crawshayi*)
Madamzebra.com
theequinest.com
arkive.org
wikipedia
pawsforwildlife.co.uk
en.wikipedia.org
africaimagelibrary.com

Plains Zebra: Upper Zambezi (*Equus burchelli zambeziensis*)
http://www.africabespoke.com/wp-content/uploads/2009/05/Lochinvar
-National-Park-150x150.jpg
http://www.jenmansafaris.com/african-countries/tanzania/attractions/saadani
-national-park.html
http://www.zambia-the-african-safari.com/images/zebras2.jpg
http://www.noble-minded.org/photography.html
http://www.sunsafaris.com/safari/zambia/kafue-national-park/
http://www.yukiba.com/7433-zambia-africa-photo.html
http://www.rainbowsafaris.biz/Kafue.html
http://www.justluxe.com/travel/luxury-vacations/feature-1428217.php

Plains Zebra: Grant's (*Equus burchelli boehmi*)
Macdonald (1984), p. 132
Madamzebra.com
treknature.com
theequinest.com
en.wikipedia.org
brightszoo.com
wikipedia.com
naturephoto-cz.com
pawsforwildlife.co.uk
johnwasserman.com
zoochat.com

Plains Zebra: Quagga (*Equus burchelli quagga*)—EXTINCT
Macdonald (1984), p. 137
Groves (1974), p. 72
http://www.quaggaproject.org/; Amsterdam specimen
http://www.quaggaproject.org/; Bamberg specimen
http://www.quaggaproject.org/; Basle specimen
http://www.quaggaproject.org/; Berlin specimen
http://www.quaggaproject.org/; Darmstadt
http://www.quaggaproject.org/; Edinburgh
http://www.quaggaproject.org/; Frankfurt
http://www.quaggaproject.org/; Kazan
http://www.quaggaproject.org/; Leiden
http://www.quaggaproject.org/; Mainz

Grevy's Zebra (*Equus grevyi*)
Macdonald (1984), p. 110
Macdonald (1984), p. 119
Macdonald (1984), p. 128
Macdonald (1984), p. 134
Hosking and Withers (1996), p. 77
wikipedia
art.com
carnivoraforum.com
commons.wikimedia.org
madamzebra.com
theequinest.com

Mountain Zebra: Hartman's (*Equus zebra hartmannae*)
Macdonald (1984), p. 111
Groves (1974), p. 71
Nowak (1999), p. 1021

Frandsen (1992), p. 144
biodiversityexplorer.org
Madamzebra.com (upper)
madamzebra.com (lower)
shoarns.com
zoochat.com
arkive.org

Mountain Zebra: Cape (*Equus zebra zebra*)
Macdonald (1984), p. 123
Macdonald (1984), p. 133
Groves (1974), p. 71
Frandsen (1992), p. 142
Madamzebra.com
biodiversityexplorer.org
south-africa-tours-and-travel.com
arkive.org

African Wild Ass: Somali (*Equus africanus somaliensis*)
sandiegozoo.org
ln.widipedia.org
arkive.org
okapia.wordpress.com
zoochat.com
en.wikipedia.org
treknature.com
biolib.cz
factzoo.com
theequinest.com

African Wild Ass: Nubian (*Equus africanus africanus*)
Kimura et al. (2010), fig. 1
http://dpc.uba.uva.nl/cgi/t/text/text-idx?c=zoomed;sid=b1971dea61cd3bc2
    aa0902085c8e9883;rgn=main;idno=m8101a06;view=text
Groves (1974), p. 107
egyptheritage.com
mustangs4us.com
animalinfo.org
biogeodb.stri.si.edu
flickr.com

African Wild Ass: Atlas (*Equus africanus atlanticus*)—EXTINCT
Kimura et al. (2010), drawing, fig. 1

Przewalski's Horse (*Equus ferus przewalskii*)
Moehlman (2002a), p. 82
en.wikipedia.org
worldwildlife.org
quantum-conservation.org
conservationcenters.org
arkive.org
britannica.hk

Kiang: Eastern (*Equus kiang holdereri*)
arkive.org
http://animaldiversity.ummz.umich.edu/site/resources/david_blank/kiang.jpg
    /view.html
cs.wikipedia.org
biolib.cz
zt.tibet.cn
etc.usf.edu

Kiang: Western (*Equus kiang kiang*)
http://wgbis.ces.iisc.ernet.in/energy/water/paper/TR123/section5.htm
http://www.asmjournals.org/doi/abs/10.1644/835.1.?journalCode=mmsp
Schaller (1998), p. 166

Kiang: Southern (*Equus kiang polyodon*)
Schaller (1998), p. 169

Asiatic Wild Ass: Mongolian Khulan (*Equus hemionus hemionus*)
arkive.org

Asiatic Wild Ass: Gobi Khulan (*Equus hemionus leuteus*)
http://reino-animal.webege.com/detalle.php?id_especie=110
http://www.cryptomundo.com/cryptozoo-news/kulan-zoo/
http://dtbook.egloos.com/9326537
http://www.equids.org/aswildass.php
http://www.zoobojnice.sk/cicavce/zivocich/kulan-turkmensky
http://www.edgeofexistence.org/mammals/species_info.php?id=14

Asiatic Wild Ass: Kulan (*Equus hemionus kulan*)
arkive.org

Asiatic Wild Ass: Onager (*Equus hemionus onager*)
Moehlman (2002a), p. 62
arkive.org

Asiatic Wild Ass: Indian (*Equus hemionus khur*)
Moehlman (2002a), p. 63
arkive.org

Asiatic Wild Ass: Syrian (*Equus hemionus hemippus*)—EXTINCT
http://en.wikipedia.org/wiki/Syrian_wild_ass
http://www.biolib.cz/en/image/id172157/

Appendix 6

Derivation of equid phylogenies

We downloaded a consensus phylogenetic tree of all seven species from the 10kTrees website (Arnold, Matthews, and Nunn 2010); a polytomy between *E. kiang*, *E. hemionus*, and *E. africanus* was resolved using the multi2di function of the APE package (Paradis, Claude and, Strimmer 2004), written in R (R Core Team 2012). This tree was used to run all species-level statistical tests and served as the base for the subspecies-level tree. We resolved the subspecies of *E. burchelli* as follows: *E. b. boehmi* was placed ancestral to *E. b. quagga* and other subspecies following (Leonard et al. 2005); *E. b. quagga* was placed ancestral to the remaining *E. burchelli* subspecies because it is often listed as a separate species (Wilson and Reeder 1993); *E. b. chapmani* and *E. b. antiquorum* were set as sister taxa because they are currently recognized as the same subspecies (Wilson and Reeder 1993), although there is disagreement as to this point. The relationship among the three subspecies of *E. africanus* is unclear because *E. a. atlanticus* is extinct; we set these subspecies as a polytomy. The validity of the three subspecies of *E. kiang* is also unclear and may just represent a cline; therefore, we set these subspecies as a polytomy. The *E. kiang* and *E. africanus* polytomies were both resolved as described above using the multi2di function. While some have placed *E. kiang* and *E. hemionus* as sister taxa (Oakenfull, Lim, and Ryder 2000; Piras et al. 2009; Steiner and Ryder 2011), we followed A. Harris and Porter (1980) and Kruger and colleagues (2005) by placing the subspecies of *E. kiang* as derived from the subspecies of *E. heminous*. Following Kruger and coworkers (2005), we placed *E. kiang* as sister to *E. h. leuteus*, *E. h. kulan* as ancestral to both *E. kiang* and *E. h. leuteus*, and *E. h. khur* as sister to *E. h. kulan*. Following A. Harris and Porter (1980) and Kruger and associates (2005), we placed *E. h. onager* as ancestral to this entire group (*E. kiang*, *E. h. leuteus*, *E. h. khur*,

E. h. kulan), and *E. h. hemippus* was placed as ancestral to all other *E. heminous* and *E. kiang* subspecies. *E. b. crawshayi*, *E. b. zambeziensis*, and *E. h. heminous* were excluded from the tree and all further analyses due to a lack of published evidence of their relationships to other subspecies. Branch lengths on the subspecies tree were initially based on those of the consensus tree, and the remaining branch lengths within each species were divided up equally.

Appendix 7

Phylogenetic analyses

As we had many more candidate factors than species or subspecies, we ran single factor phylogenetic generalized least squares (PGLS) analyses (gls function in the nlme module [Pinheiro et al. 2012], with a correlation structure based on the phylogenetic tree) for every explanatory factor (all percentage range overlap by tabanid activity [twelve versions, one through twelve consecutive months]), Old World average monthly maximum temperature values reaching 25°C and so on for each additional degree up to 30°C, woodland habitat, tsetse fly range, spotted hyena range, lion range, mean group size (species level only), and maximum group size (species level only) against the following quantitative measures of coloration at the species and subspecies level: number of face stripes, number of neck stripes, flank striping, rump striping, shadow stripes, belly stripe number, and leg striping intensity. For species-level analyses, given the small number of species (N = 7), we could only test a maximum of four factors simultaneously in one model. We chose the four factors with the lowest p-values from the individual PGLS tests and included them in an AICc model selection procedure (Burnham and Andersen 2002). To test as many hypotheses as possible, we only allowed one (the lowest p-value) tabanid score, one (the lowest p-value) temperature score, one predator (hyena or lion) score, or one group size score into these final analyses, even if other versions still had lower p-values than factors that did enter the model. Using the MuMIn module (K. Barton 2013) in R, we ran all possible AIC models, including these four factors (K. Barton 2013). Summary statistics from the all-possible-models procedure are presented for each factor, and these are weighted by their importance to the best models relative to the worst models (K. Barton 2013). In Tables 8.1 and 8.2, their importance scores (I: the sum of the model weights that each factor appears in) ranges from 0 to 1, where 0–0.2 = relatively unimportant, 0.2–0.5 = moderately important, and 0.5–1.0 = very important. Procedures were identical for subspecies-level tests, except that AICc models included six factors: the best tabanid score, the best temperature score, tsetse fly score, hyena score, lion score, and woodland score. For dichotomous coloration scores (presence of stripes [Y/N] at both the species and subspecies level, presence of shadow stripes [Y/N] at the species level), we ran phylogenetically corrected logistic regression models (PLR) using the compar.gee function in APE. We ran single factor models of all predictor variables similar to the PGLS analyses above, reporting the results of those individual tests. We reconstructed the evolutionary history of leg stripe intensity using the getAncStates function of the geiger module (Harmon et al. 2009) in R.

References

Aiello, S.E. 2002–2012. *The Merck veterinary manual online*. Merck Sharp & Dohme Corp. Accessed 25 April 2013. http://www.merckmanuals.com/vet/index.html.

Alatalo, R.V., and J. Mappes. 1996. Tracking the evolution of warning signals. *Nature* 382:485–503.

Albert, A., and J.A. Anderson. 1984. On the existence of maximum likelihood estimates in logistic regression models. *Biometrika* 71:1–10.

Allan, S.A., J.F. Day, and J.D. Edman. 1987. Visual ecology of biting flies. *Ann Rev Entomol* 32:297–316.

Allan, S.A., and J.G. Stoffolano. 1986a. Effects of background contrast on visual attraction and orientation of *Tabanus nigrovittatus* (Diptera: Tabanidae). *Environ Entomol* 15:689–694.

———. 1986b. The effects of hue and intensity on visual attraction of adult *Tabanus nigrovittatus* (Diptera: Tabanidae). *J Med Entomol* 23:83–91.

Allen, W.L., R. Baddeley, I.C. Cuthill, and N.E. Scott-Samuel. 2012. A quantitative test of the predicted relationship between countershading and lighting environment. *Am Nat* 180:762–776.

Allen, W.L., R. Baddeley, N.E. Scott-Samuel, and I.C. Cuthill. 2013. The evolution and function of pattern diversity in snakes. *Behav Ecol* 24:1237–1250.

Allen, W.L., M. Stevens, and J.P. Higham. 2014. Character displacement of Cercopithecini primate visual signals. *Nat Commun* 5. doi:10.1038.

Allsopp, R. 1972. The role of game animals in the maintenance of endemic and enzootic trypanosomiasis in the Lambwe valley, south Nyanza District, Kenya. *Bull World Health Organ* 47:735–746.

Alverson, D.R., and R. Noblet. 1977. Activity of female tabanidae

(Diptera) in relation to selected meteorological factors in South Carolina. *J Med Entomol* 14:197–200.

Anderson, J.R., and G.R. DeFoliart. 1961. Feeding behavior and host preferences of some black flies (Diptera: Simuliidae) in Wisconsin. *Ann Entomol Soc Am* 54:716–729.

Anderson, N.E., J. Mubanga, E.M. Fevre, K. Picozzi, M.C. Eisler, R. Thomas, and S.C. Welburn. 2011. Characterization of the wildlife reservoir for human and animal trypanosomiasis in the Luangwa Valley, Zambia. *PLOS Negl Trop Dis* 5:e1211.

Arnold, C., L.J. Matthews, and C.L. Nunn. 2010. The 10kTrees website: a new online resource for primate phylogeny. *Evol Anthropol* 19:114–118.

Asa, C.S. 2002. Equid reproductive biology. In *Equids: zebras, asses and horses, status survey and conservation action plan*, ed. P.D. Moehlman, 113–117. Gland, Switzerland: IUCN.

Auty, H., A. Mundy, R.D. Fyumagwa, K. Picozzi, S. Welburn, and R. Hoare. 2008. Health management of horses under high challenge from trypanosomes: a case study from Serengeti, Tanzania. *Vet Parasitol* 154:233–241.

Bahloul, K., O.B. Pereladova, N. Soldatova, G. Fisenko, E. Sidorenko, and A.J. Sempere. 2001. Social organization and dispersion of introduced kulans (*Equus hemionus kulan*) and Przewalski horses (*Equus przewalskii*) in the Bukhara Reserve, Uzbekistan. *J Arid Environ* 47:309–323.

Baker, R.R., and G.A. Parker. 1979. The evolution of bird colouration. *Philos Trans R Soc Lond B Biol Sci* 287:63–130.

Baldacchino, F., J. Cadier, A. Porciani, B. Buatois, L. Dormont, and P. Jay-Robert. 2013. Behavioural and electrophysiological responses of females of two species of tabanid to volatiles in urine of different mammals. *Med Vet Entomol* 27:77–85.

Baldacchino, F., S. Manon, L. Peuch, B. Buatois, L. Dormont, and P. Jay-Robert. 2014. Olfactory and behavioural responses of tabanid horseflies to octenol, phenols and aged horse urine. *Med Vet Entomol* 28:201–209.

Baldacchino, F., L. Puech, S. Manon, L.R. Hertzog, and P. Jay-Robert. 2014. Biting behaviour of Tabanidae on cattle in mountainous summer pastures, Pyrenees, France, and effects of weather variables. *Bull Entomol Res* 104:471–479.

Baldwin, H.A. 1971. Instrumentation for remote observation of physiology and behavior. *Proc Symp Biometry CSIR*, Pretoria, 150–156.

Barass, R. 1960. The settling of tsetse flies *Glossina morsitans* Westwood (Diptera, Muscidae) on cloth screens. *Entomologia Experimentalis et Applicata* 3:59–67.

Bard, J.B.L. 1977. Unity underlying different zebra striping patterns. *J Zool Lond* 183:527–539.

———. 1981. A model for generating aspects of zebra and other mammalian coat patterns. *J Theoret Biol* 93:363–385.

Barton, K. 2013. MuMIn: Multi-model inference. R package version 1.9.0. http://CRAN.R-project.org/package=MuMIn.

Barton, R. 1985. Grooming site preferences in primates and their functional implications. *Int J Primatol* 6:519–531.

Basile, M., A. Boivin, A. Boutin, C. Blois-Heulin, M. Hausberger, and A. Lemasson. 2009. Socially dependent auditory laterality in domestic horses (*Equus caballus*). *Anim Cognit* 12:611–619.

Bauer, I.E., J. McMorrow, and D.W. Yalden. 1994. The historic ranges of three equid species in north-east Africa: a quantitative comparison on environmental tolerances. *J Biogeogr* 21:169–182.

Baylis, M. 1996. Effect of defensive behavior by cattle on the feeding success and nutritional state of the tsetse fly *G. pallidipes* (Diptera: Glossinidae). *Bull Entomol Res* 86:329–336.

Beddard, F.E. 1892. *Animal coloration: an account of the principal facts and theories*. London: Swan Sonnenschein.

Behrens, R.R. 2002. *False colours: art, design and modern camouflage*. Dysart, IA: Bobolink Books.

Benesch, A.R., and S. Hilsberg. 2003. Infrarot-thermographische Untersuchungen der Oberflachentemperatur bei Zebras. *Zool Garten NF* 2, S:74–82.

Benesch, A.R., and S. Hilsberg-Merz. 2006. Oberflachen-temperaturen bei Zebrastreifen. *Nature und Museum* 136 (3–4):49–56.

Bennett, D.K. 1980. Stripes do not a zebra make. Part I, A cladistics analysis of *Equus*. *Syst Zool* 29:272–287.

Berger, J. 1986. *Wild horses of the great basin*. Chicago: Chicago University Press.

Bertram, B.C.R. 1973. Sleeping sickness survey in the Serengeti area (Tanzania), 1971. III. Discussion of the relevance of the trypanosome survey to the biology of large mammals in the Serengeti. *Acta Tropica* 30:36–47.

Bhasin, A., A.J. Mordue (Luntz), and W. Mordue. 2000. Responses of the biting midge *Culicoides impunctatus* to acetone, CO_2 and 1-octen-3-ol in a wind tunnel. *Med Vet Entomol* 14:300–307.

Bidlingmayer, W.L. 1994. How mosquitoes see traps: role of visual responses. *J Amer Mosquito Control Assoc* 10:272–279.

Bidlingmayer, W.L., and D.G. Hem. 1980. The range of visual attraction and the effect of competitive visual attractants upon mosquito (Diptera: Culicidae) flight. *Bull Entomol Res* 70:321–342.

Blaho, M., A. Egri, L. Bahidszki, G. Kriska, R. Hegedus, S. Akesson, and G. Horvath. 2012. Spottier targets are less attractive to tabanid flies: on the tabanid-repellency of spotty fur patterns. *PLOS ONE* 7:e41138.

Blaho, M., A. Egri, D. Szaz, G. Kriska, S. Akesson, and G. Horvath. 2013. Stripes disrupt odour attractiveness to biting horseflies: battle between ammonia, CO_2, and colour pattern for dominance in the sensory systems of host-seeking tabanids. *Physiol Behav* 119:168–174.

Bökönyi, S. 1974. *The Przevalsky horse*. London: Soubenier Press.

Boorman, J. 1993. Biting midges (Ceratopogonidae). In *Medical insects and arachnids*, eds. R.P. Lane and R.W. Crosskey, 288–309. London: Chapman & Hall.

Borst, A. 2014. Fly visual course control: behaviour, algorithms, and circuits. *Nature Rev Neurosci* 15:590–599.

Bouskila, A., S. Renan, E. Speyer, D. Ben-Natan, I. Zaibel, and S. Bar-David. 2013. Group composition and behavior of *Equus hemionus* near a water source in the Negev Desert. *J Vet Behav* 8:e1–e25.

Boyd, L. 1988. "The behavior of Przewalski's horses." PhD diss., Cornell University.

———. 1991. The behavior of Przewalski's horses and its importance to their management. *Appl Anim Behav Sci* 29:301–318.

Boyd, L., and K.A. Houpt. 1994. Activity patterns. In *Przewalski's horse: the history and biology of an endangered species*, eds. L. Boyd and K.A. Houpt, 195–227. Albany: State University of New York Press.

Bracken, G.K., W. Hanec, and A.J. Thorsteinson. 1962. The orientation of horse flies and deer flies (Tabanidae: Diptera). II. The role of some visual factors in the attractiveness of decoy silhouettes. *Can J Zool* 40:685–695.

Brady, J. 1972. The visual responsiveness of the tsetse fly *Glossina morsitans* Westw. (Glossinidae) to moving objects: the effects of hunger, sex, host odour and stimulus characteristics. *Bull Entomol Res* 62:257–279.

Brady, J., and W. Shereni. 1988. Landing responses of the tsetse fly *Glossina morsitans morsitans* Westwood and the stable fly *Stomoxys calcitrans* (L.) (Diptera: Glossinidae and Muscidae) to black-and-white patterns: a laboratory study. *Bull Entomol Res* 78:301–311.

Britten, K.H., T.D. Thatcher, and T. Caro. 2016. Zebras and biting flies: quantitative analysis of reflected light from zebra coats in their natural habitat. *PLOS ONE* 11:e0154504.

Broom, M., and G.D. Ruxton. 2005. You can run—or you can hide: optimal strategies for cryptic prey against pursuit predators. *Behav Ecol* 16:534–540.

Brown, C.C. 1992. African horse sickness: An Olympic risk? *Compendium on Continuing Education for the Practicing Veterinarian* 14:239–241.

Browne, S.M., and G.F. Bennett. 1980. Color and shape as mediators of host-seeking responses of simuliids and tabanids (Diptera) in the Tantramar marshes, New Brunswick, Canada. *J Med Entomol* 17:58–62.

Burchell, W.J. 1824. *Travels in the interior of southern Africa*. Vol. 2. London: Longman, Hurst, Rees, Orme and Brown.

Burnett, A.M., and K.L. Hays. 1974. Some influences of meteorological factors on flight activity of female horse flies (Diptera: Tabanidae). *Environ Entomol* 3:515–521.

Burnham, K.P., and D. Andersen. 2002. *Model selection and multi-model inference*. 2nd ed. New York: Springer.

Burridge, M.J.M., H.W. Reid, N.B. Pullan, R.W. Sutherst, and E.B. Wain. 1970. Survey for trypanosome infections in domestic cattle and wild animals in areas of East Africa. II. Salivarian trypanosome infections in wild animals in Busoga District, Uganda. *Brit Veterin J* 126:627–633.

Bursell, E. 1984. Effects of host odour on the behavior of tsetse. *Insect Sci Applic* 5:345–349.

Burtt, E.H. 1981. The adaptiveness of animal colors. *BioScience* 31:723–729.

Bush, G.L., S.M. Case, A.C. Wilson, and J.L. Patton. 1977. Rapid speciation and chromosomal evolution in mammals. *Proc Natl Acad Sci USA* 74:3942–3946.

Carbone, C., G.M. Mace, S.C. Roberts, and D.W. Macdonald. 1999. Energetic constraints on the diet of terrestrial carnivores. *Nature* 402:286–288.

Caro, T. 1994. Cheetahs of the Serengeti Plains: group living in an asocial species. Chicago: University of Chicago Press.

———. 1999a. Abundance and distribution of mammals in Katavi National Park, Tanzania. *Afr J Ecol* 37:305–313.

———. 1999b. Conservation monitoring: estimating mammal densities in woodland habitats. *Anim Conserv* 2:305–315.

———. 1999c. Densities of mammals in partially protected areas: the Katavi ecosystem of western Tanzania. *J Appl Ecol* 36:205–217.

———. 2005. *Antipredator defenses in birds and mammals*. Chicago: University of Chicago Press.

———. 2008. Decline of large mammals in the Katavi-Rukwa ecosystem of western Tanzania. *Afr Zool* 43:99–116.

———. 2009. Contrasting coloration in terrestrial mammals. *Philos Trans R Soc Lond B Biol Sci* 364:537–48.

———. 2011. Black and white coloration in mammals: review and synthesis. In *Animal camouflage: mechanisms and function*, ed. M. Stevens and S. Merilaita, 298–329. Cambridge: Cambridge University Press.

———. 2013. The colours of extant mammals. *Sem Cell Dev Biol* 24:542–552.

———. In press. Kingdon on colouration: crested rats, guenons and zebras. *J E Afr Nat Hist*.

Caro, T., C.M. Graham, C.J. Stoner, and J.K. Vargas. 2004. Adaptive significance of antipredator behaviour in artiodactyls. *Anim Behav* 67:205–228.

Caro, T., A. Izzo, R.C. Reiner, H. Walker, and T. Stankowich. 2014. The function of zebra stripes. *Nat Commun* 5:3535.

Caro, T., and C. Melville. 2012. Investigating colouration in large and rare mammals: the case of the giant anteater. *Ethol Ecol Evol* 24:104–115.

Caro, T., and T. Stankowich. 2015. Concordance on zebra stripes: a comment on Larison *et al.* (2015). *R Soc Open Sci* 2:150323.

Cena, K., and J.A. Clark. 1973. Thermographic measurements of the surface temperatures of animals. *J Mammal* 54:1003–1007.

Cena, K., and J.L. Monteith. 1975. Transfer processes in animal coats: I. Radiative transfer. *Proc R Soc Lond B Biol Sci* 188:413–423.

Chainey, J.E. 1993. Horse-flies, deer-flies and clegs (Tabanidae). In *Medical insects and arachnids*, eds. R.P. Lane and R.W. Crosskey, 310–332. London: Chapman & Hall.

Challier, A. 1982. The ecology of tsetse (*Glossina* spp.) (Diptera, Glossinidae): a review (1970–1981). *Insect Sci Applic* 3:97–143.

Challier, A., and C. Laveissiere. 1973. Un nouveau piege pour la capture des glossines (*Glossina*: Diptera, Muscidae): description et essais sur le terrain. *Cah ORSTOM, Ser Entomol med Parasitol* 11:251–262.

Cheng, T-H. 1958. The effect of biting fly control on weight gain in beef cattle. *J Econ Entomol* 51:275–278.

Chvala, M., L. Lyneborg, and J. Moucha. 1972. *The horse flies of Europe—Diptera, Tabanidae.* Copenhagen: Entomological Society of Copenhagen.

Clausen, P.H., I. Adeyemi, B. Bauer, M. Breloeer, F. Salchow, and C. Staak. 1998. Host preferences of tsetse (Diptera: Glossinidae) based on bloodmeal identifications. *Med Vet Entomol* 12:169–180.

Clements, H.S., C.J. Tambling, M.W. Hayward, and G.I.H. Kerley. 2015. An objective approach to determining the weight ranges of prey preferred by and accessible to the five large African carnivores. *PLOS ONE* 9:e101054.

Cloudsley-Thompson, J.L. 1969. *The zoology of tropical Africa.* London: Weidenfeld and Nicolson.

———. 1980. *Tooth and claw: defensive strategies in the animal world.* London: J.M. Dent & Sons.

———. 1984. How the zebra got its stripes: new solutions to an old problem. *Biologist* 31:226–228.

———. 1999. Multiple factors in the evolution of animal coloration. *Naturwissenschaften* 86:123–132.

Clutton-Brock, T.H., P.J. Greenwood, and R.P. Powell. 1976. Ranks and relationships in Highland ponies and Highland cows. *Z Tierpsychol* 41:202–216.

Cott, H.B. 1940. *Adaptive colouration in animals.* London: Methuen.

———. 1946–1947. The edibility of birds: illustrated by five years' experiments and observations (1941–1946) on the food preferences of the hornet, cat and man: and considered with special reference to the theories of adaptive coloration. *Proc Zool Soc Lond* 116:371–524.

———. 1966. *Colouration in animals.* London: Methuen.

Cotton, D. 1998. The curious thing about quaggas. *Biologist* 45:175–176.

Creel, S., and N.M. Creel. 2002. *The African wild dog: biology, ecology, and conservation.* Princeton, NJ: Princeton University Press.

Creel, S., P. Schuette, and D. Christianson. 2014. Effects of predation risk on group size, vigilance, and foraging behavior in an African ungulate community. *Behav Ecol* 25:773–784.

Cresswell, W. 1994. Flocking is an effective anti-predation strategy in redshanks, *Tringa totanus. Anim Behav* 47:433–442.

Crosskey, R.W. 1993a. Blackflies (Simuliidae). In *Medical insects and arachnids*, eds. R.P. Lane and R.W. Crosskey, 241–287. London: Chapman & Hall.

———. 1993b. Stable-flies and horn-flies (bloodsucking Muscidae). In *Medical insects and arachnids*, eds. R.P. Lane and R.W. Crosskey, 389–428. London: Chapman & Hall.

Crowell-Davis, S.L., K.A. Houpt, and C.M. Carini. 1986. Mutual grooming and nearest-neighbor relationships among foals of *Equus caballus. Appl Anim Behav Sci* 15:113–123.

Cullen, A. 1969. *Window onto wilderness*. Nairobi: East African Publishing House.

Cuthill, I.C., M. Stevens, J. Sheppard, T. Maddocks, C.A. Parraga, and T.S. Troscianko. 2005. Disruptive coloration and background pattern matching. *Nature* 434:72–74.

Darst, C.R., M.E. Cummings, and D.C. Cannatella. 2006. A mechanism for diversity in warning signals: conspicuousness versus toxicity in poison frogs. *Proc Nat Acad Sci* 103:5852–5857.

Darwin, C.R. 1871. *The descent of man, and selection in relation to sex*. Vol. 2. London: John Murray.

———. 1906. *The descent of man, and selection in relation to sex*. 2nd ed. London: John Murray.

de Andrade Silva, M.A., and J. Marques da Silva. 1959. On the incidence of trypanosomiasis in game. *Anais Inst Med Trop* 16:209–212.

Dillman, J.S.S., and A.J. Townsend. 1979. A trypanosomiasis survey of wild animals in the Luangwa valley, Zambia. *Acta Tropica* 36:349–356.

Dorward, D.C., and A.I. Payne. 1975. Deforestation, the decline of the horse, and the spread of the tsetse fly and trypanosomiasis (nagana) in nineteenth century Sierra Leone. *J Afr Hist* 16:239–256.

Downes, C.M., J.B. Theberge, and S.M. Smith. 1986. The influence of insects on the distribution, microhabitat choice, and behavior of the Burwash caribou herd. *Can J Zool* 64:622–629.

Drager, N., and D. Mehlitz. 1978. Investigations on the prevalence of trypanosome carriers and the antibody response in wildlife in northern Botswana. *Tropenmedizin und Parasitologie* 29:223–233.

Dransfield, R.D. 1984. The range of attraction of the biconical trap for *Glossina pallidipes* and *G. brevipalpis*. *Insect Sci Applic* 5:363–368.

Dumbacher, J.P., and R.C. Fleischer. 2001. Phylogenetic evidence for colour pattern convergence in toxic pitohuis: Mullerian mimicry in birds? *Proc R Soc Lond B Biol Sci* 268:1971–1976.

Duncan, P., and P. Cowtan. 1980. An unusual choice of habitat helps Camargue horses to avoid blood-sucking flies. *Biol Behav* 5:55–60.

Duncan, P., and C. Groves. 2013. Genus Equus: asses, zebras. In *Mammals of Africa*, vol. 5, *Carnivores, pangolins, equids and rhinoceroses*, eds. J. Kingdon and M. Hoffmann, 412–413. London: Bloomsbury.

Duncan, P., and N. Vigne. 1970. The effect of group size in horses on the rate of attacks by blood-sucking flies. *Anim Behav* 27:623–625.

Edmunds, M. 1974. *Defence in animals*. New York: Longman.

Egri, A., M. Blaho, G. Kriska, R. Farkas, M. Gyurkovszky, S. Akesson, and G. Horvath. 2012. Polarotactic tabanids find striped patterns with brightness and/or polarization modulation least attractive: an advantage of zebra stripes. *J Exp Biol* 215:736–745.

Egri, A., M. Blaho, A. Sandor, G. Kriska, M. Gyurkovszky, R. Farkas, and G. Horvath. 2012. New kind of polarotaxis governed by degree of polarization:

attraction of tabanid flies to differently polarizing host animals and water surfaces. *Naturwissenschaften* 99:407–416.

Eilam, D. 2005. Die hard: a blend of freezing and fleeing as a dynamic defense— implications for the control of defensive behavior. *Neurosci Biobehav Rev* 29:1181–1191.

Elliott, J.P., I. McTaggart Cowan, and C.S. Holling. 1977. Prey capture by the African lion. *Can J Zool* 55:1811–1828.

Eltringham, S.K. 1979. *The ecology and conservation of large mammals.* Basingstoke, UK: Macmillan.

Endler, J.A. 1978. A predator's view of animal color patterns. *Evol Biol* 11:319–364.

Endler, J.A., and J. Mappes. 2004. Predator mixes and the conspicuousness of aposematic signals. *Am Nat* 163:532–547.

Erdsack, N., G. Dehnhardt, and W. Hanke. 2013. Coping with heat: function of the natal coat of Cape fur seal (*Arctocephalus pusillus pusillus*) pups in maintaining core body temperature. *PLOS ONE* 8:e72081.

Espmark, Y., and R. Langvatn. 1979. Lying down as a means of reducing fly harassment in red deer (*Cervus elaphus*). *Behav Ecol Sociobiol* 5:51–55.

Estes, R.D. 1991. *The behavior guide to African mammals.* Berkeley: University of California Press.

———. 1993. *The safari companion: a guide to watching African mammals.* White River Junction, VT: Chelsea Green Publishing.

Factsheets: Equine Encephalitis. 2009. Center for Food Security and Public Health, Iowa State University. Accessed 25 April 2013. http://www .cfsph.iastate.edu/Factsheets/pdfs/easter_wester_venezuelan_equine _encephalomyelitis.pdf.

Factsheets: Equine Infectious Anemia. 2009. Center for Food Security and Public Health, Iowa State University. Accessed 25 April 2013. http://www.cfsph .iastate.edu/Factsheets/pdfs/equine_infectious_anemia.pdf.

Feh, C., T. Boldsukh, and C. Tourenq. 1994. Are family groups in equids a response to cooperative hunting by predators? the case of Mongolian khulans (*Equus hemionus luteus* Matschie). *Rev Ecol* 49:11–20.

Feh, C., and J. de Mazieres. 1993. Grooming at a preferred site reduces heart rate in horses. *Anim Behav* 46:1191–1194.

Feh, C., B. Munkhtuya, S. Enkhbold, and T. Sukhbaatar. 2001. Ecology and social structure of the Gobi khulan *Equus hemionus* subsp. in the Gobi B National Park, Mongolia. *Biol Cons* 101:51–61.

Feh, C., N. Shah, M. Rowen, R. Reading, and S.P. Goyal. 2002. Status and action plan for the Asiatic wild ass (*Equus hemionus*). In *Equids: zebras, asses and horses, status survey and conservation action plan*, ed. P.D. Moehlman, 62–71. Gland, Switzerland: IUCN.

Finch, V.A., I.L. Bennett, and C.R. Holmes. 1984. Coat colour in cattle: effect on thermal balance, behavior and growth, and relationship with coat type. *J Agric Sci Camb* 102:141–147.

Finch, V.A., R. Dmi'el, R. Bowman, A. Shkolnik, and C.R. Taylor. 1980. Why black coats in hot deserts? Effects of coat colour on heat exchanges of wild and domestic goats. *Physiol Zool* 350:19–25.

Fischhoff, I.R., S.R. Sundaresan, J. Cordingley, and D.I. Rubenstein. 2007. Habitat use and movements of plains zebra (*Equus burchelli*) in response to predation danger from lions. *Behav Ecol* 18:725–729.

FitzGibbon, C.D. 1989. A cost to individuals with reduced vigilance in groups of Thomson's gazelles hunted by cheetahs. *Anim Behav* 37:508–510.

Fletcher, T., C. Janis, and E. Rayfield. 2009. Finite element analysis of ungulate jaws: can mode of digestive physiology be distinguished by jaw robusticity in extant and extinct ungulates? *J Vert Paleontol* 27:96A–97A.

Foil, L. 1989. Tabanids as vectors of disease agents. *Parasitology Today* 5:88–96.

Foil, L., and J.A. Hogsette. 1994. Biology and control of tabanids, stable flies and horn flies. *Rev Sci Tech Off Int Epiz* 13:1125–1158.

Foil, L., D. Stage, W.V. Adams, M.S. Issel, and C.J. Issel. 1985. Observations of tabanid feeding on mares and foals. *Am J Vet Res* 46:1111–1113.

Fontes, J., and R. Fontes. 2002. *How the zebra got its stripes: tales from around the world*. New York: Random House.

Foster, G., H. Krijger, and S. Bangay. 2007. Zebra fingerprints: towards a computer-aided identification system for individual zebra. *Afr J Ecol* 45:225–257.

Frandsen, S. 1992. *Southern Africa's mammals: a field guide*. 2nd ed. Sandton, South Africa: Frandsen Publishers.

Fraser, S., A. Callahan, D. Klassen, and T.N. Sherratt. 2007. Empirical tests of the role of disruptive coloration in reducing detectability. *Proc R Soc Lond B Biol Sci* 274:1325–1331.

Fuller, A., S.K. Maloney, P.R. Kamerman, G. Mitchell, and D. Mitchell. 2000. Absence of selective brain cooling in free-ranging zebras in their natural habitat. *Exper Physiol* 85:209–217.

Funston, P.J., M.G.L. Mills, and H.C. Biggs. 2001. Factors affecting the hunting success of male and female lions in the Kruger National Park. *J Zool* 253:419–431.

Galton, F. 1851. *South Africa*. Minerva Library.

Garland, T., Jr. 1983. The relation between maximal running speed and body mass in terrestrial mammals. *J Zool* 199:157–170.

Geisbauer, G., U. Griebel, A. Schmid, and B. Timney. 2004. Brightness discrimination and neutral point testing in the horse. *Can J Zool* 82:660–670.

Gibson, D.O. 1974. Batesian mimicry without distastefulness. *Nature* 250:77–79.

———. 1980. The role of escape in mimicry and polymorphism. I. The response of captive birds to artificial prey. *Biol J Linn Soc* 14:201–214.

Gibson, G. 1992. Do tsetse flies "see" zebras? a field study of the visual response of tsetse to striped targets. *Physiol Entomol* 17:141–147.

Gibson, G., and S.J. Torr. 1999. Visual and olfactory responses of haematophagous Diptera to host stimuli. *Med Vet Entomol* 13:2–23.

Gibson, G., and S. Young. 1991. The optics of tsetse flies in relation to their behaviour and ecology. *Physiol Entomol* 16:273–282.

Gillies, M.T. 1980. The role of carbon dioxide in host-finding by mosquitoes (Diptera: Culicidae): a review. *Bull Entomol Res* 70:525–532.

Ginsberg, J.R., and D.I. Rubenstein. 1990. Sperm competition and variation in zebra mating behavior. *Behav Ecol Sociobiol* 26:427–434.

Gittleman, J.L., and P.H. Harvey. 1980. Why are distasteful prey not cryptic? *Nature* 286:149–150.

Glasgow, J.P. 1946. The seasonal abundance of blood-sucking flies in a grassed woodland area in central Tanganyika. *J Anim Ecol* 15:93–103.

———. 1963. *The distribution and abundance of tsetse*. New York: Macmillan.

———. 1967. Recent fundamental work on tsetse flies. *Ann Rev Entomol* 12:421–438.

Gochfeld, M. 1981. Responses of black skimmers to high-intensity distress notes. *Anim Behav* 29:1137–1145.

Godfrey, D., J.N. Lythgoe, and D.A. Rumball. 1987. Zebra stripes and tiger stripes: the spatial frequency distribution of the pattern compared to that of the background is significant in display and crypsis. *Biol J Linn Soc* 32:427–433.

Gotmark, F. 1994. Are bright birds distasteful? A re-analysis of H.B. Cott's data on the edibility of birds. *J Avian Biol* 25:184–197.

Grange, S., P. Duncan, J-M. Gaillard, A.R.E. Sinclair, P.J.P. Gogan, C. Packer, H. Hofer, and M. East. 2004. What limits the Serengeti zebra population? *Oecologia* 140:523–532.

Grannett, P., and E.J. Hansens. 1956. The effect of biting fly control on milk production. *J Econ Entomol* 49:465–467.

Greaves, N. 1988. *When hippo was hairy and other tales from Africa*. Cape Town: Struik Publishers.

Green, C.H. 1986. Effects of colours and synthetic odours on the attraction of *Glossina pallidipes* and *G. morsitans morsitans* to traps and screens. *Physiol Entomol* 11:411–421.

———. 1988. The effect of colour on trap- and screen-orientated responses in *Glossina palpalis palpalis* (Robineau-Desvoidy) (Diptera: Glossinidae). *Bull Entomol Res* 78:591–604.

Green, C.H., and S. Flint. 1986. An analysis of colour effects in the performance of the F2 trap against *Glossina pallidipes* Austen and *G. morsitans morsitans* Westwood (Diptera: Glossinidae). *Bull Entomol Res* 76:409–418.

Griffith, S.C., T.H. Parker, and V.A. Olson. 2006. Melanin- versus carotenoid-based sexual signals: is the difference really so black and red? *Anim Behav* 71:749–763.

Groves, C. 1974. *Horses, asses and zebras in the wild*. Newton Abbott, UK: David and Charles.

———. 2002. Taxonomy of living equidae. In *Equids: zebras, asses and horses, status survey and conservation action plan*, ed. P.D. Moehlman, 94–107. Gland, Switzerland: IUCN.

———. 2013. Family Equidae: asses, zebras. In *Mammals of Africa*, vol. 5, *Carnivores, pangolins, equids and rhinoceroses*, eds. J. Kingdon and M. Hoffmann, 410–412. London: Bloomsbury.

Groves, C., and C.H. Bell. 2004. New investigations on the taxonomy of the zebras genus *Equus*, subgenus *Hippotigris*. *Mamm Biol* 69:182–196.

Grubb, P. 1981. *Equus burchelli. Mammalian Species* 157:1–9.

Hack, M.A., R. East, and D.I. Rubenstein. 2002. Status and action plan for the plains zebra (*Equus burchelli*). In *Equids: zebras, asses and horses, status survey and conservation action plan*, ed. P.D. Moehlman, 43–60. Gland, Switzerland: IUCN.

Hadithi, M., and A. Kennaway. 1984. *Greedy zebra*. London: Hodder Headline.

Hagman, M., and A. Forsman. 2003. Correlated evolution of conspicuous coloration and body size in poison frogs (Dendrobatidae). *Evolution* 57:2904–2910.

Hall, J.R., I.C. Cuthill, R. Baddeley, A.S. Attwood, M.R. Munafò, and N.E. Scott-Samuel. 2016. Dynamic dazzle distorts speed perception. *PLOS ONE* 11:e0155162.

Hall, J.R., I.C. Cuthill, R. Baddeley, A.J. Shohet, and N.E. Scott-Samuel. 2013. Camouflage, detection and identification of moving targets. *Proc R Soc Lond B Biol Sci* 280:20130064.

Hamilton, W.J., III. 1973. *Life's color code*. New York: McGraw-Hill.

Hamilton, W.J., III., and F. Heppner. 1967. Radiant solar energy and the function of black homeotherm pigmentation: an hypothesis. *Science* 155:196–197.

Hammack, L., and L.S. Hesler. 1996. Phenylpropanoids as attractants for adult *Stomoxys calcitrans* (Diptera: Muscidae). *J Med Entomol* 33:859–862.

Hanggi, E.B., J.F. Ingersoll, and T.L. Waggoner. 2007. Color vision in horses (*Equus caballus*): deficiencies identified using a pseudoisochromatic plate test. *J Comp Physiol* 121:65–72.

Hardie, R.C. 1984. Properties of photoreceptors R7 and R8 in dorsal marginal ommatidia in the compound eyes of *Musca* and *Calliphora*. *J Comp Physiol A* 154:157–165.

———. 1986. The photoreceptor array of the dipteran retina. *Trends Neurosci* 9:419–423.

Hardie, R.C., N. Franceschini, and N. McIntyre. 1979. Electrophysiological analysis of fly retina. II. Spectral and polarization sensitivity in R7 and R8. *J Comp Physiol* 133:23–39.

Hardie, R.C., K. Vogt, and A. Rudolph. 1989. The compound eye of the tsetse fly (*Glossina morsitans morsitans* and *Glossina palpalis palpalis*). *J Insect Physiol* 35:423–431.

Harmon, L., J. Weir, C. Brock, R. Glor, W. Challenger, and G. Hunt. 2009. geiger: analysis of evolutionary diversification. R package version 1.3-1. http://CRAN .Rproject.org/package=geiger.

Harris, A.H., and L.S.W. Porter. 1980. Late Pleistocene horses of Dry Cave, Eddy County, New Mexico. *J Mammal* 61:46–65.

Harris, R.H.T.P. 1930. *Report on the bionomics of the tsetse fly*. Peitermaritzburg, South Africa: Provincial Administration of Natal.

Harrison, H. 1940. *The game experiment. Tsetse research report, 1935–1938.* Tanganyika Territory Dept. of Tsetse Research. Dar es Salaam: Govt. Printer.

Hauglund, K., S.B. Hagen, and H.M. Lampe. 2006. Responses of domestic chicks (*Gallus gallus domesticus*) to multimodal aposematic signals. *Behav Ecol* 17:392–398.

Hayward, M.W. 2006. Prey preferences of the spotted hyaena (*Crocuta crocuta*) and degree of dietary overlap with the lion (*Panthera leo*). *J Zool Lond* 270: 606–614.

Hayward, M.W., G.J. Hayward, C.J. Tambling, and G.I.H. Kerley. 2011. Do lions *Panthera leo* actively select prey or do prey preferences simply reflect chance responses via evolutionary adaptations to optimal foraging? *PLOS ONE* 6:e23607.

Hayward, M.W., P. Henschel, J. O'Brien, M. Hofmeyr, G. Balme, and G.I.H. Kerley. 2006. Prey preferences of the leopard (*Panthera pardus*). *J Zool Lond* 270: 298–313.

Hayward, M.W., M. Hofmeyr, J. O'Brien, and G.I.H. Kerley. 2006. Prey preferences of the cheetah (*Acinonyx jubatus*) (Felidae: Carnivora): morphological limitations or the need to capture rapidly consumable prey before kleptoparasites arrive? *J Zool Lond* 270:615–627.

Hayward, M.W., and G.I.H. Kerley. 2005. Prey preferences of the lion (*Panthera leo*). *J Zool Lond* 267:309–322.

Hayward, M.W., J. O'Brien, M. Hofmeyr, and G.I.H. Kerley. 2006. Prey preferences of the African wild dog *Lycaon pictus* (Canidae: Carnivora): ecological requirements for conservation. *J Mammal* 87:1122–1131.

Hayward, M.W., and R. Slotow. 2009. Temporal partitioning of activity in large African carnivores: tests of multiple hypotheses. *S Afr J Wildl Res* 39:109–125.

Helin, S., N. Ohtaishi, and L. Houji. 1999. *The mammals of China.* Beijing: China Forestry Publishing House.

Helmholtz, H. v. (1867) 1962. *Treatise on Physiological Optics.* Vol. 3. New York: Dover, 1962. English translation by J.P.C. Southall for the Optical Society of America (1925) from the 3rd German ed. of *Handbuch der physiologischen Optik*, first published in 1867 (Leipzig: Voss).

Hemmer, H. 1990. *Domestication: the decline of environmental appreciation.* Cambridge: Cambridge University Press.

Hetem, R.S., B.A. de Witt, L.G. Fick, A. Fuller, G.I.H. Kerley, L.C.R. Meyer, D. Mitchell, and S.K. Maloney. 2009. Body temperature, thermoregulatory behaviour and pelt characteristics of three colour morphs of springbok (*Antidorcas marsupialis*). *Comp Biochem Physiol A* 152:379–388.

Hetem, R.S., W.S. Strauss, B.G. Heusinkveld, S. de Bie, H.H.T. Prins, and S.E. van Wieren. 2011. Energy advantages of orientation to solar radiation in three African ruminants. *J Therm Biol* 36:452–460.

Higuchi, R., B. Bowman, M. Freiberger, O.A. Ryder, and A.C. Wilson. 1984. DNA sequences from the quagga, an extinct member of the horse family. *Nature* 312:282–284.

Hingston, R.W.G. 1933. *The meaning of animal colour and adornment*. London: Edward Arnold.

Hollander, A.L., and R.E. Wright. 1980. Impact of tabanids on cattle: blood meal size and preferred feeding sites. *J Econ Entomol* 73:431–433.

Holloway, M.T.P., and R.J. Phelps. 1991. The responses of *Stomoxys* spp. (Diptera: Muscidae) to traps and artificial host odours in the field. *Bull Entomol Res* 81:51–55.

Hongo, A., and M. Akimoto. 2003. The role of incisors in selective grazing by cattle and horses. *J Agric Science* 140:469–477.

Hornok, S., G. Foldvari, E. Vilmos, V. Naranjo, R. Farkas, and J. de la Fuente. 2008. Molecular identification of *Anaplasma marginale* and rickettsial endosymbionts in blood-sucking flies (Diptera: Tabanidae, Muscidae) and hard ticks (Acari: Ixodidae). *Vet Parasitol* 154:354–359.

Horvath, G., ed. 2014. *Polarized light and polarization in animal sciences*. Berlin: Springer-Verlag.

Horvath, G., M. Blaho, G. Kriska, R. Hegedus, B. Gerics, R. Farkas, and S. Akesson. 2010. An unexpected advantage of whiteness in horses: the most horsefly-proof horse has a depolarizing white coat. *Proc R Soc Lond B Biol Sci* 277:1643–1650.

Horvath, G., and Z. Csabai. 2014. Polarization vision of aquatic insects. In *Polarized light and polarization in animal sciences*, ed. G. Horvath, 113–145. Berlin: Springer-Verlag.

Horvath, G., J. Majer, L. Horvath, I. Szivak, and G. Kriska. 2008. Ventral polarization vision in tabanids: horseflies and deerflies (Diptera: Tabanidae) are attracted to horizontally polarized light. *Naturwissenschaften* 95:1093–1100.

Horvath, G., and D. Varju. 2004. *Polarized light in animal vision: polarization patterns in nature*. Berlin: Springer-Verlag.

Hosking, D., and M.B. Withers. 1996. *Larger animals of East Africa*. London: Harper Collins.

Houston, A.I., M. Stevens, and I.C. Cuthill. 2007. Animal camouflage: compromise or specialize in a 2 patch-type environment? *Behav Ecol* 18:769–775.

How, M.J., and J.M. Zanker. 2014. Motion dazzle induced by zebra stripes. *Zoology* 117:163–170.

Hribar, L.J., D.J. Leprince, and L.D. Foil. 1991. Design for a canopy trap for collecting horse flies (Diptera: Tabanidae). *J Am Mosq Contr Assoc* 7:657–659.

Hughes, A.E., R.S. Major-Elliott, and M. Stevens. 2015. The role of stripe orientation in target capture success. *Front Zool* 12:17.

Hughes, A.E., J. Troscianko, and M. Stevens. 2014. Motion dazzle and the effects of target patterning on capture success. *BMC Biology* 14:201. http://www.biomedcentral.com/1471-2148/14/201.

Humphries, D.A., and P.M. Driver. 1967. Erratic display as a device against predators. *Science* 156:1767.

Hutchinson, J.C.D., and G.D. Brown. 1969. Penetrance of cattle coats by radiation. *J Appl Physiol* 26:454–464.

Ioannou, C.S., and J. Krause. 2009. Interactions between background matching and motion during visual detection can explain why cryptic animals keep still. *Biol Lett* 5:191–193.

Ioannou, C.S., C.R. Tosh, L. Neville, and J. Krause. 2007. The confusion effect—from neural networks to reduced predation risk. *Behav Ecol* 19:126–130.

Ishida, N., T. Oyunsuren, S. Mashima, H. Mukoyama, and N. Saitou. 1995. Mitochondrial DNA sequences of various species of the genus *Equus* with special reference to the phylogenetic relationship between Przewalskii's wild horse and domestic horse. *J Mol Evol* 41:180–188.

Jackson, J.F., W. Ingram, and H.W. Campbell. 1976. The dorsal pigmentation pattern of snakes as an antipredator strategy: a multivariate approach. *Am Nat* 110:1029–1053.

Jacobs, J. 1974. Quantitative measurements of food selection. *Oecologia* 14:413–417.

Janis, C.M., and D. Ehrhardt. 1988. Correlation of relative muzzle width and relative incisor width with dietary preference in ungulates. *Zool J Linn Soc* 92:267–284.

Jeschke, J.M., and R. Tollrian. 2007. Prey swarming: which predators become confused and why? *Anim Behav* 74:387–393.

Jessen, C., R. Dmi'el, I. Choshniak, D. Ezra, and G. Kuhnen. 1998. Effects of dehydration and rehydration on body temperatures in the black Bedouin goat. *Pflugers Archiv* 436:659–666.

Jobling, B. n.d. Anatomical drawings of biting flies. *British Museum*, 102–104.

Jones, K.A., A.L. Jackson, and G.D. Ruxton. 2011. Prey jitters: protean behaviour in grouped prey. *Behav Ecol* 22:831–836.

Jordan, A.M. 1976. Tsetse flies as vectors of trypanosomes. *Vet Parasitol* 2:143–152.

———. 1993. Tsetse-flies (*Glossinidae*). In *Medical insects and arachnids*, eds. R.P. Lane and R.W. Crosskey, 333–388. London: Chapman & Hall.

Jordan, A.M., F. Lee-Jones, and B. Weitz. 1962. The natural hosts of tsetse flies in northern Nigeria. *Annals Trop Med Parasitol* 56:430–442.

Joubert, E. 1972. Activity patterns shown by mountain zebra *Equus zebra hartmannae* in South West Africa with reference to climatic factors. *Zool Africana* 7:309–331.

Kaczensky, P., O. Ganbaatar, H. von Wehrden, and C. Walzer. 2008. Resource selection by sympatric wild equids in the Mongolian Gobi. *J Appl Ecol* 45:1762–1769.

Kangwagye, T.N. 1977. Reaction of large mammals to biting flies in Rwenzori National Park, Uganda. In *Proceedings of the First East African Conference on Entomology and Pest Control 1976*, 32–44. Nairobi.

Kappmeier, K., and E.M. Nevill. 1999. Evaluation of coloured targets for the attraction of *Glossina brevipalis* and *Glossina austeni* (Diptera: Glossinidae) in South Africa. *Onderstepoort J Vet Res* 66:291–305.

Keiper, R., and J. Berger. 1982. Refuge seeking pest avoidance by feral horses in desert and island environments. *Appl Anim Ethol* 9:111–120.

Keiper, R., and H. Receveur. 1992. Social interactions of free-ranging Przewalski horses in semi-reserves in the Netherlands. *Appl Anim Behav Sci* 33:303–318.

Kelley, J.L., J.L. Fitzpatrick, and S. Merilaita. 2013. Spots and stripes: ecology and colour pattern evolution in butterflies. *Proc R Soc Lond B Biol Sci* 280:20122730.

Keymer, I.F. 1969. A survey of trypanosome infections in wild ungulates in the Luangwa valley, Zambia. *Annals Trop Med Parasitol* 63:195–200.

Kikula, I.S. 1980. *Landsat satellite data for vegetation mapping in Tanzania: the case of the Rukwa Region*. Bureau of Resource Assessment and Land Use Planning, Research Report No. 41, University of Dar es Salaam.

Kimura, R. 1998. Mutual grooming and preferred associate relationships in a band of free-ranging horses. *Appl Anim Behav Sci* 59:265–276.

———. 2000. Relationship of the type of social organization to scent-marking and mutual-grooming behaviour in Grevy's (*Equus grevyi*) and Grant's zebras (*Equus burchelli bohmi*). *J Equine Sci* 11:91–98.

Kimura, B., F.B. Marshall, S. Chen, S., Rosenbom, P.D. Moehlman, N. Tuross, R.C. Sabin, J. Peters, B. Barich, H. Yohannes, F. Kebede, R. Teclai, A. Beja-Pereira, and C.J. Mulligan. 2010. Ancient DNA from Nubian and Somali wild ass provides insights into donkey ancestry and domestication. *Proc R Soc Lond B Biol Sci* 278:50–57.

King, S.R.B., and J. Gurnell. 2010. Effects of fly disturbance on the behaviour of a population of reintroduced Przewalski horses (*Equus ferus przewalskii*) in Mongolia. *Appl Anim Behav Sci* 125:22–29.

Kingdon, J. 1979. *East African mammals: an atlas of evolution in Africa*. Vol. 3, part B. 1st ed. Chicago: University of Chicago Press.

———. 1984. The zebra's stripes: an aid to group cohesion. In *The encyclopedia of mammals*, ed. D. Macdonald, 486–487. Oxford: Equinox.

———. 1997. *The Kingdon field guide to African mammals*. London: Academic Press.

Kinghorn, A., and W. Yorke. 1912a. Further observations on the trypanosomes of game and livestock in northeastern Rhodesia. *Annals Trop Med Parasitol* 6:483–493.

———. 1912b. Trypanosomes obtained by feeding wild *Glossina morsitans* on monkeys in Luangwa valley, Northern Rhodesia. *Annals Trop Med Parasitol* 6:317–325.

Kipling, R. 1902. *Just so stories for little children*. Oxford: Oxford University Press.

Kirk-Spriggs, A.H. 2012. Dedication: the life, career and major achievements of Brian Roy Stuckenberg (1930–2009). *African Invertebrates* 53:1–34.

Klimov, V.V. 1988. Spatial-ethological organization of the herd of Przewalski horses (*Equus przewalskii*) in Askania-Nova. *Appl Anim Behav Sci* 21:99–115.

Klingel, H. 1969. The social organization and population ecology of the plains zebra. *Zool Africana* 4:249–263.

———. 1974. A comparison of the social behaviour of the Equidae. Calgary Ungulate Conference, IUCN Publication, pp. 124–132.

———. 1975. Social organization and reproduction in equids. Supplement, *J Reprod Fertil* 23:7–11.

———. 1977. Observations on social organization and behaviour of African and Asiatic wild asses (*Equus africanus* and *Equus hemionus*). *Z Tierpsychol* 44: 323–331.

———. 2013. *Equus quagga*, plains zebra (common zebra). In *Mammals of Africa*. Vol. 5, *Carnivores, pangolins, equids and rhinoceroses*, eds. J. Kingdon and M. Hoffmann, 428–437. London: Bloomsbury.

Komoin-Oka, C., P. Truc, Z. Bengaly, P. Formenty, G. Duvallet, F. Lauginie, J.P. Raath, A.E. N'Depo, and Y. Leforban. 1994. Etude de la prévalence des infections à trypanosomes chez différentes espèces d'animaux sauvages du parc national de la Comoé en Côte d'Ivoire: résultats préliminaires sur la comparaison de trois méthodes de diagnostic. *Revue d'élevage et de médecine vétérinaire des pays tropicaux* 47:189–194.

Krakauer, D.C. 1995. Groups confuse predators by exploiting perceptual bottlenecks: a connectionist model of the confusion effect. *Behav Ecol Sociobiol* 36:421–429.

Krause, J., and G.D. Ruxton. 2002. *Living in groups*. Oxford: Oxford University Press.

Krenn, H.W., and A. Horst. 2012. Form, function and evolution of the mouthparts of blood-feeding Arthropods. *Arthropod Struct Develop* 41:101–118.

Krinsky, W.L. 1976. Animal disease agents transmitted by horse flies and deer flies (Diptera, Tabanidae). *J Med Entomol* 13:225–275.

Kriska, G., B. Bernath, R. Farkas, and G. Horvath. 2009. Degrees of polarization of reflected light eliciting polarotaxis in dragonflies (Odonata), mayflies (Ephemeroptera) and tabanid flies (Tabanidae). *J Insect Physiol* 55:1167–1173.

Kriticos, D.J., B.L. Webber, A. Leriche, N. Ota, I. Macadam, J. Bathols, and J.K. Scott. 2011. CliMond: global high resolution historical and future scenario climate surfaces for bioclimatic modeling. *Methods in Ecology & Evolution* 3:53–64.

Kruger, K., C. Gaillard, G. Stranzinger, and S. Rieder. 2005. Phylogenetic analysis and species allocation of individual equids using microsatellite data. Supplement 1, *J Anim Breed Genet* 122:78–86.

Kruuk, H. 1972. *The spotted hyaena: a study of predation and social behavior*. Chicago: University of Chicago Press.

Kuhnen, G. 1997. Selective brain cooling reduces respiratory water loss during heat stress. *Comp Biochem Physiol A* 118:891–895.

Land, M.F., and D-E. Nilsson. 2012. *Animal eyes*. Oxford: Oxford University Press.

Lane, R.P. 1993. Sandflies (*Phlebotominae*). In *Medical insects and arachnids*, eds. R.P. Lane and R.W. Crosskey, 78–119. London: Chapman & Hall.

Lane, R.P., and R.W. Crosskey, eds. 1993. *Medical insects and arachnids*. London: Chapman & Hall.

Larison, B., R.J. Harrigan, D.I. Rubenstein, and T.B. Smith. 2015. Concordance on zebra stripes is not black and white: response to comment by Caro & Stanko-wich 2015. *R Soc Open Sci* 2:150329.

Larison, B., R.J. Harrigan, H.A. Thomassen, D.I. Rubenstein, A.M. Chan-Golston, E. Li, and T.B. Smith. 2015. How the zebra got its stripes: a problem with too many solutions. *R Soc Open Sci* 2:140452.

Lariviere, S., and F. Messier. 1996. Aposematic behaviour in the striped skunk, *Mephitis mephitis*. *Ethology* 102:986–992.

Lechocinski, N., and S. Breugnot. 2011. Fiber orientation measurement using polarizing imaging. *J Cosmet Sci* 62:85–100.

Lehane, M. 2005. *The biology of blood-sucking in insects*. 2nd ed. Cambridge: Cambridge University Press.

Leimar, O., M. Enquist, and B. Sillen-Tullberg. 1986. Evolutionary stability of aposematic coloration and prey unprofitability: a theoretical analysis. *Am Nat* 128:469–490.

Lemasson, A., A. Boutin, S. Boivin, C. Blois-Heulin, and M. Hausberger. 2009. Horse (*Equus caballus*) whinnies: a source of information. *Anim Cognit* 12:693–704.

Lembo, T., K. Hampson, H. Auty, C.A. Beesley, P. Bessell, C. Packer, J. Halliday, R. Fyumagwa, R. Hoare, E. Ernest, C. Mentzel, T. Mlengeya, K. Stamey, P.P. Wilkins, and S. Cleaveland. 2011. Serological surveillance of anthrax in the Serengeti ecosystem, Tanzania, 1996–2009. *Emerging Insect Diseases* 17: 387–394.

Leonard, J.A., N. Rohland, S. Glaberman, R.C. Fleischer, A. Caccone, and M. Hofreiter. 2005. A rapid loss of stripes: the evolutionary history of the extinct quagga. *Biol Lett* 1:291–295.

Linklater, W.L. 2000. Adaptive explanation in socio-ecology: lessons from the Equidae. *Biol Rev* 75:1–20.

Ljetoff, M., I. Folstad, F. Skarstein, and N.G. Yoccoz. 2007. Zebra stripes as an amplifier of individual quality? *Ann Zool Fennici* 44:368–376.

Louw, G.N. 1993. *Physiological animal ecology*. Harlow, Essex, UK: Longman Scientific and Technical.

Lowenstein, J.M., and O.A. Ryder. 1984. Immunological systematics of the extinct quagga (Equidae). *Experimentia* 41:1192–1193.

Ludwig, A., M. Pruvost, M. Reissmann, N. Benecke, G.A. Brockmann, P. Castanos, M. Cieslak, S. Lippold, L. Llorente, A-S. Malaspinas, M. Slatkin, and M. Hofreiter. 2009. Coat color variation at the beginning of horse domestication. *Science* 324:485.

Lydekker, R. 1912. *The horse and its relatives*. London: George Allen.

Lythgoe, J.N. 1979. *The ecology of vision*. Oxford: Clarendon Press.

MacClintock, D. 1976. *A natural history of zebras*. New York: Charles Scribner's Sons.

Macdonald, D.W., ed. 1984. *The encyclopedia of mammals*. London: Allen & Unwin.

Mackerras, I.M. 1954. The classification and distribution of Tabanidae (Diptera). I. General review. *Austral J Zool* 2:431–454.

Maddock, L. 1979. The "migration" and grazing succession. In *Serengeti: dynamics of an ecosystem*, eds. A.R.E. Sinclair and M. Norton Griffiths, 104–129. Chicago: University of Chicago Press.

Maia, A.S.C., R.G. da Silva, and E.C.A. Bertipaglia. 2005. Environmental and genetic variation of the effective radiative properties of the coat of Holstein cows under tropical conditions. *Livestock Prod Sci* 92:307–315.

Malcolm, J.R., and H. van Lawick. 1975. Notes on wild dogs (*Lycaon pictus*) hunting zebras. *Mammalia* 39:231–240.

Malele, I.I., S.M. Kinung'hi, H.S. Nyingilili, L.E. Matemba, J.K. Sahani, T.D.K. Mlengeya, M. Wambura, and S.N. Kibona. 2007. *Glossina* dynamics in and around the sleeping sickness endemic Serengeti ecosystem of northwestern Tanzania. *J Vect Biol* 32:263–268.

Mangani, B. 1962. Buffalo kills lion. *Afr Wild Life* 16:27.

Mappes, J., N. Marples, and J.A. Endler. 2005. The complex business of survival by aposematism. *Trends Ecol Evol* 20:598–603.

Marshall, N.J. 2000. Communication and camouflage with the same "bright" colours in reef fishes. *Philos Trans R Soc Lond B Biol Sci* 355:1243–1248.

Matthews, L.H. 1971. *The life of mammals*. Vol. 2. New York: Universe Books.

McComb, K., C. Packer, and A. Pusey. 1994. Roaring and numerical assessment in contests between groups of female lions, *Panthera leo*. *Anim Behav* 47:379–387.

McLeod, D.N.K. 1987. Zebra stripes. *New Scientist* 115:68.

Melin, A.D., D.W. Kline, C. Hiramatsu, and T. Caro. 2016. Zebra stripes through the eyes of their predators, zebras, and humans. *PLOS ONE* 11:e0145679.

Merilaita, S. 1998. Crypsis through disruptive coloration in an isopod. *Proc R Soc Lond B Biol Sci* 265:1059–1064.

Merilaita, S., and V. Kaitala. 2002. Community structure and the evolution of aposematic coloration. *Ecol Lett* 5:495–501.

Merilaita, S., and J. Lind. 2005. Background-matching and disruptive coloration, and the evolution of cryptic coloration. *Proc R Soc Lond B Biol Sci* 272: 665–670.

Merilaita, S., A. Lyytinen, and J. Mappes. 2001. Selection for cryptic coloration in a visually heterogeneous habitat. *Proc R Soc Lond B Biol Sci* 268:1925–1929.

Merilaita, S., and G.D. Ruxton. 2007. Aposematic signals and the relationship between conspicuousness and distinctiveness. *J Theoret Biol* 245:268–277.

Milinski, M. 1984. A predator's cost of overcoming the confusion-effect of swarming prey. *Anim Behav* 32:1157–1162.

Miller, L.A. 1951. Observations of the bionomics of some northern species of Tabanidae (Diptera). *Can J Zool* 29:240–263.

Mills, M.G., and L.B. Patterson. 2009. Not just black and white: pigment pattern development and evolution in vertebrates. *Sem Cell Dev Biol* 20:72–81.

Milner J., and D. Hewitt. 1969. Weight of horses: improved estimates based on girth and length. *Can Vet J* 10:314–316.

Moehlman, P.D. 1985. The odd-toed ungulates: order Perissodactyla. In *Social odours in mammals*, vol. 2, eds. R.E. Brown and D.W. Macdonald. Oxford: Clarendon Press.

———. 1998. Feral asses (*Equus africanus*): intraspecific variation in social organization in arid and mesic habitats. *Appl Anim Behav Sci* 60:171–195.

———. 2002a. *Status survey and conservation action plan. Equids: zebras, asses and horses*. Gland, Switzerland: IUCN/SSC Equid Specialist Group.

———. 2002b. Status and action plan for the African wild ass (*Equus africanus*). In *Equids: zebras, asses and horses, status survey and conservation action plan*, ed. P.D. Moehlman, 2–10. Gland, Switzerland: IUCN.

———. 2002c. List of equids on the 2002 IUCN Red List of Threatened Species. In *Equids: zebras, asses and horses, status survey and conservation action plan*, ed. P.D. Moehlman, 163. Gland, Switzerland: IUCN.

Moehlman, P.D., F. Kebede, and H. Yohannes. 2013. *Equus africanus*, African wild ass (Somali wild ass). In *Mammals of Africa*, vol. 5, *Carnivores, pangolins, equids and rhinoceroses*, eds. J. Kingdon and M. Hoffmann, 414–417. London: Bloomsbury.

Moen, A.N., and C.W. Severinghaus. 1984. Hair depths of the winter coat of white-tailed deer. *J Mammal* 65:497–499.

Mohamed-Ahmed, M.M., and S. Mihok. 2009. Alighting of Tabanidae and muscids on natural and simulated hosts in the Sudan. *Bull Entomol Res* 99: 561–571.

Mohr, E. 1971. *The Asiatic wild horse*. London: J.A. Allen.

Moloo, S.K. 1993. The distribution of *Glossina* species in Africa and their natural hosts. *Insect Sci Applic* 14:511–527.

Molyneux, D.H., and R.W. Ashford. 1983. *The biology of trypanosoma and leishmania, parasites of man and domestic animals*. London: Taylor and Francis.

Moodley, Y., and E.H. Harley. 2005. Population structuring in mountain zebras (*Equus zebra*): the molecular consequences of divergent demographic histories. *Conserv Genet* 6:953–968.

Morris, D. 1990. *Animal watching: a field guide to animal behaviour*. London: Johnathan Cape.

Mostert, C. 2012. *How the zebra got its stripes*. © Cari Mostert, Wildmoz.

Mottram, J.C. 1915. Some observations on pattern-blending with reference to obliterative shading and concealment of outline. *Proc Zoo Soc Lond* 85: 679–692.

———. 1916. An experimental determination of the factors which cause patterns to appear conspicuous in nature. *Proc Zool Soc Lond* 86:383–419.

Mulla, A.F., and L.R. Rickman. 1988. The isolation of human serum-resistant *Trypanosoma* (*Trypanozoon*) species from zebra and impala in Luangwa valley, Zambia. *Trans R Soc Trop Med Hyg* 82:718.

Mullens, B.A., and R.R. Gerhardt. 1979. Feeding behavior of some Tennessee tabanidae. *Environ Entomol* 8:1047–1051.

Muoira, P.K., P. Muruthi, K. Waititu, A. Boru, D.M. Hassan, and N.O. Oguge. 2007.

Anthrax outbreak among Grevy's zebra (*Equus grevyi*) in Samburu, Kenya. *Afr J Ecol* 45:483–489.

Murray, J.D. 2007. *Mathematical biology. I: An introduction*. 3rd ed. New York: Springer.

Muturi, C.N., J.O. Ouma, I.I. Malele, R.M. Ngure, J.J. Rutto, K.M. Mithofer, J. Enyaru, and D.K. Masiga. 2011. Tracking the feeding patterns of tsetse flies (*Glossina* genus) by analysis of bloodmeals using mitochondrial cytochromes genes. *PLOS ONE* 6:e17284.

Muzari, M.O., L.F. Skerratt, R.E. Jones, and T.L. Duran. 2010. Alighting and feeding behavior of tabanid flies on horses, kangaroos and pigs. *Vet Parasitol* 170: 104–111.

Mwambu, P.M., and M.H. Woodford. 1972. Trypanosomes from game animals of the Queen Elizabeth National Park, western Uganda. *Trop Anim Health Prod* 4:152–155.

Napier Bax, S. 1937. The senses of smell and sight in *Glossina swynnertoni*. *Bull Entomol Res* 28:539–582.

Ndegwa, P.N., and S. Mihok. 1999. Development of odour-baited traps for *Glossina swynnertoni* (Diptera: Glossinidae). *Bull Entomol Res* 89:255–261.

Neill, S.R. StJ., and J.M. Cullen. 1974. Experiments on whether schooling by their prey affects the hunting behaviour of cephalopods and fish predators. *J Zool* 172:549–569.

Nilsson, A.N., ed. 1997. *Aquatic insects of North Europe: a taxonomic handbook*. Stendstrup, Denmark: Apollo Books.

Njiru, Z.K., J.N. Makumi, S. Okoth, J.M. Ndungu, and W.C. Gibson. 2004. Identification of trypanosomiasis in *Glossina pallidipes* and *G. longipennis* in Kenya. *Infection, Genetics and Evolution* 4:29–35.

Norton-Griffiths, M. 1978. *Counting animals: techniques currently used in African wildlife ecology*. Publication No. 1. Nairobi: African Wildlife Foundation.

Novellie, P., M. Lindeque, P. Lindeque, P. Lloyd, and J. Koen. 2002. Status and action plan for the mountain zebra (*Equus zebra*). In *Equids: zebras, asses and horses, status survey and conservation action plan*, ed. P.D. Moehlman, 28–42. Gland, Switzerland: IUCN.

Nowak, R.M. 1999. *Walker's mammals of the world*. 6th ed. Baltimore: Johns Hopkins University Press.

Oakenfull, E.A., and J.B. Clegg. 1998. Phylogenetic relationships within the genus *Equus* and the evolution of α and θ globin genes. *J Mol Evol* 47:772–783.

Oakenfull, E.A., H.N. Lim, and O.A. Ryder. 2000. A survey of equid mitochondrial DNA: implications for the evolution, genetic diversity and conservation of *Equus*. *Conserv Genet* 1:341–355.

Olsen, S.L. 2006. Early horse domestication on the Eurasian steppe. In *Documenting domestication: new genetic and archeological paradigms*, eds. M.A. Zeder, D.G. Bradley, E. Emshwiller, and B.D. Smith, 245–269. Los Angeles: University of California Press.

Oristland, N.A. 1970. Energetic significance of absorption of solar radiation in polar homeotherms. In *Antarctic Ecology*, vol. 1, ed. M.W. Holgate, 464–470. New York: Academic Press.

Orlando, L., A. Ginolhac, G. Zhang, D. Froese, A. Albrechtsen, M. Stiller, M. Schubert, E. Cappellini, B. Petersen, I. Moltke, P.L.F. Johnson, M. Fumagalli, J.Y. Vilstrup, M. Raghavan, T. Korneliussen, A-S. Malaspinas, J. Vogt, D. Szklarczyk, C.D. Kelstrup, J. Vinther, A. Dolocan, J. Stenderup, A.M.V. Velazquez, J. Cahill, M. Rasmussen, X. Wang, J. Min, G.D. Zazula, A. Seguin-Orlando, C. Mortensen, K. Magnussen, J.F. Thompson, J. Weinstock, K. Gregersen, K.H. Roed, V. Eisenmann, C.J. Rubin, D.C. Miller, D.F. Antczak, M.F. Bertelsen, S. Brunak, K.A.S. Al-Rasheid, O. Tryder, L. Andersson, J. Mundy, A. Krogh, M.T. Gilbert, K. Kjaer, T. Sicheritz-Ponten, L.J. Jensen, J.V. Olsen, M. Hofreiter, R. Nielsen, B. Shapiro, J. Wang, and E. Willerslev. 2013. Recalibrating Equus evolution using the genome sequence of an early middle Pleistocene horse. *Nature* 499:74–78.

Orlando, L., J.L. Metcalf, M.T. Alberdi, M. Telles-Antunes, D. Bonjean, M. Otte, F. Martin, V. Eisenmann, M. Mashkour, F. Morello, J.L. Prado, R. Salas-Gismondi, B.J. Shockey, P.J. Wrinn, S.K. Vasil'ev, N.D. Ovodov, M.I. Cherry, B. Hopwood, D. Male, J.J. Austin, C. Haani, and A. Cooper. 2009. Revising the recent evolutionary history of equids using ancient DNA. *Proc Natl Acad Sci USA* 106:21754–21759.

Owen-Smith, N., and M.G.L. Mills. 2008. Predator-prey size relationships in an African large-mammal food web. *J Anim Ecol* 77:173–183.

Pan, Y.S. 1964. Variation in hair characters over the body in Sahiwal zebu and Jersey cattle. *Aust J Agric Res* 15:346–356.

Paradis, E., J. Claude, and K. Strimmer. 2004. APE: analyses of phylogenetics and evolution in R language (ver. 3.0-1). *Bioinformatics* 20:289–290.

Parsons, R., C. Aldous-Mycock, and M.R. Perrin. 2007. A genetic index for stripe-pattern reduction in the zebra: the quagga project. *S Afr J Wildl Res* 37: 105–116.

Paynter, Q., and J. Brady. 1993. Flight responses of tsetse flies (*Glossina*) to octenol and acetone vapour in a wind-tunnel. *Physiol Entomol* 18:102–108.

Pennacchio, O., I.C. Cuthill, P.G. Lovell, G.D. Ruxton, and J.M. Harris. 2015. Orientation to the sun by animals and its interaction with crypsis. *Funct Ecol* 29:1165–1177.

Pennacchio, O., P.G. Lovell, G. Ruxton, I.C. Cuthill, and J.W. Harris. 2015. Three-dimensional camouflage: exploiting photons to conceal form. *Am Nat* 186:553–563.

Penzhorn, B. 1984. A long-term study of social organization and behavior of Cape mountain zebras, *Equus zebra zebra*. *Z Tierpsychol* 64:97–146.

———. 1988. *Equus zebra*. *Mammalian species* 314:1–7.

———. 2013. *Equus zebra*, mountain zebra. In *Mammals of Africa*, vol. 5, *Carnivores, pangolins, equids and rhinoceroses*, eds. J. Kingdon and M. Hoffman, 438–443. London: Bloomsbury.

Penzhorn, B., and P.A. Novellie. 1991. Some behavioural traits of Cape mountain zebras (*Equus zebra zebra*) and their implications for the management of a small conservation area. *Appl Anim Behav Sci* 29:293–299.

Peterson, J.C.B. 1972. An identification system for zebras (*Equus burchelli* Gray). *E Afr Wildl J* 10:59–63.

Phelps, R.J., and M.T.P. Holloway. 1990. Alighting sites of female Tabanidae (Diptera) at Rekomitjie, Zimbabwe. *Med Vet Entomol* 4:349–356.

Phelps, R.J., and G.A. Vale. 1976. Studies on the local distribution and on the methods of host location of some Rhodesian Tabanidae (Diptera). *J Ent Soc S Afr* 39:67–81.

Pienaar, U. de. 1969. Predator-prey relations amongst the larger mammals of Kruger National Park. *Koedoe* 12:108–176.

Pinchbeck, G.L., L.J. Morrison, A. Tait, J. Langford, L. Meehan, S. Jallow, J. Jallow, A. Jallow, and R.M. Christley. 2008. Trypanosomiasis in The Gambia: prevalence in working horses and donkeys detected by whole genome amplification and PCR, and evidence for interactions between trypanosome species. *BMC Vet Res* 4:7.

Pinheiro, J., D. Bates, S. DebRoy, D. Sarkar, and the R Development Core Team. 2012. nlme: Linear and Nonlinear Mixed Effects Models. R package version 3.1–103.

Piras, F.M., S.G. Nergadze, V. Poletto, F. Cerutti, O.A. Ryder, T. Leeb, and E. Giulotto. 2009. Phylogeny of horse chromosome 5q in the genus *Equus* and centromere repositioning. *Cytogenet Genome Res* 126:165–172.

Pisciotti, I.S., and D.E. Miranda. 2005. Horse flies (DIPTERA: TABANIDAE) from the "Parque Nacional Natural Chiribiquete," Caqueta, Colombia. *Industrial University of Santander School of Biology*, 1–104.

Plotz, R.D., and W.L. Linklater. 2009. Black rhinoceros (*Diceros bicornis*) calf succumbs after lion predation attempt: implications for conservation management. *Afr Zool* 44:283–287.

Policht, R., A. Karadzos, and D. Frynta. 2011. Comparative analysis of long-range calls in equid stallions (Equidae): are acoustic parameters related to social organization? *Afr Zool* 46:18–26.

Popowics, T.E., and M. Fortelius. 1997. On the cutting edge: tooth blade sharpness in herbivorous and faunivorous mammals. *Ann Zool Fennici* 34:73–88.

Poulton, E.B. 1890. *The colours of animals: their meaning and use. Especially considered in the case of insects.* London: Kegan Paul, Trench Trübner.

———. 1902. The meaning of the white undersides of animals. *Nature* 65:596–597.

Pratt, D.J., P.J. Greenway, and M.D. Gwynne. 1966. A classification of the East African rangeland. *J Appl Ecol* 3:369–382.

Proops, L., K. McComb, and D. Reby. 2009. Cross-modal individual recognition in domestic horses (*Equus caballus*). *Proc Nat Acad Sci* 106:947–951.

Prothero, D.R., and R.M. Schoch. 2003. *Horns, tusks, and flippers: the evolution of hoofed mammals.* Baltimore: Johns Hopkins University Press.

Prudic, K.L., A.K. Skemp, and D.R. Papaj. 2006. Aposematic coloration, luminance contrast, and the benefits of conspicuousness. *Behav Ecol* 18:41–46.

Pruvost, M., R. Bellone, N. Benecke, E. Sandoval-Castellanos, M. Cieslak, T. Kuznetsova, A. Morales-Muniz, T. O'Connor, M. Reissmann, M. Hofreiter, and A. Ludwig. 2011. Genotypes of predomestic horses match phenotypes painted in Paleolithic works of cave art. *Proc Nat Acad Sci* 46:18626–18630.

Radcliffe, R.M., and S.A. Osofsky. 2002. Disease concerns for wild Equids. In *Equids: zebras, asses and horses, status survey and conservation action plan*, ed. P.D. Moehlman, 124–153. Gland, Switzerland: IUCN.

Ralley, W.E., T.D. Galloway, and G.H. Crow. 1993. Individual and group behaviour of pastured cattle in response to attack by biting flies. *Can J Zool* 71: 725–734.

Rau, R.E. 1974. Revised list of the preserved material of the extinct Cape colony quagga and notes on the relationship and distribution of southern plains zebras. *Ann S Afr Mus* 77:27–45.

R Core Team. 2012. R: a language and environment for statistical computing. Vienna, Austria: R Foundation for Statistical Computing. Retrieved from http://www.R-project.org.

Reading, R.P., H.M. Mix, B. Lhagvasuren, C. Feh, D.P. Kane, S. Dulamtseren, and S. Enkhbold. 2001. Status and distribution of khulan (*Equus hemionus*) in Mongolia. *J Zool Lond* 254:381–389.

Riggs, K. 2014. *Amazing animals: zebras*. Mankato, MN: Creative Education.

Riha, J., J. Minar, O. Kralik, and V. Krupa. 1983. Economic importance of protecting draft horses used in forestry against blood-sucking dipterous insects. *Vet Med (Praha)* 28:169–175.

Rippi, M., R.V. Alatalo, L. Lindstrom, and J. Mappes. 2001. Multiple benefits of gregariousness cover detectability costs in aposematic aggregations. *Nature* 413:512–514.

Robert, N., C. Walzer, S. Ruegg, P. Kaczensky, O. Ganbaatar, and C. Stauffer. 2005. Pathological findings in reintroduced Przewalski's horses (*Equus caballus przewalskii*) in southwestern Mongolia. *J Zoo Wildl Med* 36:273–285.

Roosevelt, T. 1910. *African game trails: an account of the wanderings of an American hunter-naturalist*. New York: Syndicate Publishing.

Roosevelt, T., and E. Heller. 1914. *Life-histories of African game animals*. Vol. 2. New York: Charles Scribner's Sons.

Roper, T.J. 1994. Conspicuousness of prey retards reversal of learned avoidance. *Oikos* 69:115–118.

Roper, T.J., and S. Redston. 1987. Conspicuousness of distasteful prey affects the strength and durability of one-trial avoidance learning. *Anim Behav* 35: 739–747.

Roth, L.S.V., A. Balkenius, and A. Kelber. 2007. Colour perception in a dichromat. *J Exp Biol* 210:2795–2800.

———. 2008. The absolute threshold of colour vision in the horse. *PLOS ONE* 11:e3711.

Rowe, C. 1999. Receiver psychology and the evolution of multicomponent signals. *Anim Behav* 58:921–931.

———. 2002. Sound improves visual discrimination learning in avian predators. *Proc R Soc Lond B Biol Sci* 269:1353–1357.

Rowland, H.M. 2008. From Abbott Thayer to the present day: what have we learned about the function of countershading? *Phil Trans R Soc B* 364: 519–527.

Rubenstein, D.I. 1986. Ecology and sociality in horses and asses. In *Ecological aspects of social evolution*, eds. D.I. Rubenstein and R.W. Wrangham, 282–302. Princeton, NJ: Princeton University Press.

———. 2006. Horses, zebras, and asses. In *The Encyclopedia of Mammals*, 2nd ed., ed. D.W. Macdonald, 690–695. New York: Facts on File.

———. 2010. Ecology, social behavior, and conservation in zebras. *Adv Study Behav* 42:231–258.

———. 2011. Family Equidae (horses and relatives). In *Handbook of Mammals of the World*. Vol. 2, *Hoofed mammals*, eds. D.E. Wilson and R.A. Mittermeier, 106–143. Barcelona, Spain: Lynx Edicions.

Rubenstein, D.I., and M.A. Hack. 1992. Horse signals: the sounds and scents of fury. *Evol Ecol* 6:254–260.

Rubenstein, D.I., and D.E. Hohmann 1989. Parasites and social behavior of island feral horses. *Oikos* 55:312–320.

Rutberg, A.T. 1987. Horse fly harassment and the social behavior of feral ponies. *Ethology* 75:145–154.

Ruxton, G.D. 2002. The possible fitness benefits of striped coat coloration for zebra. *Mamm Rev* 32:237–244.

Ruxton, G.D., A.L. Jackson, and C.R. Tosh. 2007. Confusion of predators does not rely on specialist coordinated behavior. *Behav Ecol* 18:590–96.

Ruxton, G.D., T.N. Sherratt, and M.P. Speed. 2004. *Avoiding attack: the evolutionary ecology of crypsis, warning signals and mimicry*. Oxford: Oxford University Press.

Sanchez, F. 2011. *How did stories*. © Frank Sanchez.

Santana, S.E., J.L. Alfaro, and M.E. Alfaro. 2012. Adaptive evolution of facial colour patterns in Neotropical primates. *Proc R Soc Lond B Biol Sci* 279:2204–2211.

Santana, S.E., J.L. Alfaro, A. Noonan, and M.E. Alfaro. 2013. Adaptive response to sociality and ecology drives the diversification of facial colour patterns in cattarhines. *Nat Commun* 4. doi:10.1038/ncomms3765.

Santer, R.D. 2014. A colour opponent model that explains tsetse fly attraction to visual baits and can be used to investigate more efficacious bait materials. *PLOS Neglected Trop Diseases* 8:e3360.

Schaefer, H.M., and N. Stobbe. 2006. Disruptive coloration provides camouflage independent of background matching. *Proc R Soc Lond B Biol Sci* 273:2427–2432.

Schaller, G.B. 1972. *The Serengeti lion: a study of predator-prey relations*. Chicago: University of Chicago Press.

———. 1998. *Wildlife of the Tibetan steppe*. Chicago: University of Chicago Press.

Scheel, D. 1993. Profitability, encounter rates, and prey choice of African lions. *Behav Ecol* 4:90–97.

Scheel, D., and C. Packer. 1991. Group hunting behavior of lions: a search for cooperation. *Anim Behav* 41:697–709.

Scheibe, K.M., K. Eichhorn, B. Kalz, W.J. Streich, and A. Scheibe. 1998. Water consumption and watering behavior of Przewalski horses (*Equus ferus przewalskii*) in a semireserve. *Zoo Biol* 17:181–192.

Schieltz, J.M., and D.I. Rubenstein. 2015. Caught between two worlds: genes and environment influence behavior of plains x Grevy's zebra hybrids in central Kenya. *Anim Behav* 106:17–26.

Schino, G., S. Scucchi, D. Maestripieri, and P.G. Turillazzi. 1988. Allogrooming as a tension-reduction mechanism: a behavioural approach. *Am J Primatol* 16:43–50.

Schulz, E., and T.M. Kaiser. 2013. Historical distribution, habitat requirements and feeding ecology of the genus Equus (Perissodactyla). *Mamm Rev* 43:111–123.

Schwind, R. 1991. Polarization vision in water insects and insects living on a moist substrate. *J Comp Physiol* 169:531–540.

Scott, J.M., F.L. Ramsey, and C.B. Kepler. 1981. Distance estimation as a variable in estimating bird numbers. In *Estimating numbers of terrestrial birds*, eds. C.J. Ralph and J.M. Scott, 334–340. Studies in Avian Biology, vol. 6. Lawrence, KS: Cooper Ornithological Society.

Scott-Samuel, N.E., R. Baddeley, C.E. Palmer, and I.C. Cuthill. 2011. Dazzle camouflage affects speed perception. *PLOS ONE* 6:e20233.

Scott-Samuel, N.E., G. Holmes, R. Baddeley, and I.C. Cuthill. 2015. Moving in groups: how density and unpredictable motion affect predation risk. *Behav Ecol Sociobiol* 69:867–872.

Seehausen, O., P.J. Mayhew, and J.J.M. Van Alphen. 1999. Evolution of colour patterns in East African cichlid fish. *J Evol Biol* 12:514–534.

Selous, F.C. 1908. *African nature notes and reminiscences*. London: Macmillan.

Service, M.W. 1993. Mosquitoes (*Culicidae*). In *Medical insects and arachnids*, eds. R.P. Lane and R.W. Crosskey, 120–240. London: Chapman & Hall.

Shah, N. 2002. Status and action plan for the kiang (*Equus kiang*). In *Equids: zebras, asses and horses, status survey and conservation action plan*, ed. P.D. Moehlman, 72–81. Gland, Switzerland: IUCN.

Sherratt, T.N. 2002. The coevolution of warning signals. *Proc R Soc Lond B Biol Sci* 269:741–746.

———. 2003. State-dependent risk-taking by predators in systems with defended prey. *Oikos* 103:93–100.

Shrader, A.M., S.M. Ferreira, and R.J. van Aarde. 2006. Digital photogrammetry and laser rangefinder techniques to measure African elephants. *S Afr J Wildl Res* 36:1–7.

Siegfried, W.R. 1990. Tail length and biting insects of ungulates. *J Mammal* 71: 75–78.

Sigurjonsdottir, H., M. van Dierendonck, S. Snorrason, and A.G. Thorhallsdottir.

2003. Social relationships in a group of horses without a mature stallion. *Behaviour* 140:783–804.

Silk, J.B., and G.R. Brown. 2008. Local resource competition and local resource enhancement shape primate birth sex ratios. *Proc R Soc Lond B Biol Sci* 275:1761–1765.

Sillen-Tullberg, B. 1985. The significance of coloration per se, independent of background, for predator avoidance of aposematic prey. *Anim Behav* 33:1382–1384.

Simpson, G.G. 1951. *Horses*. Oxford: Oxford University Press.

Skinner, J.D., and C.T. Chimimba. 2005. *The mammals of the southern African sub-region*. Cambridge: Cambridge University Press.

Slominski, A., J. Wortsman, P.M. Plonka, K.U. Schallreuter, R. Paus, and D.J. Tobin. 2004. Hair follicle pigmentation. *J Investig Dermatol* 124:13–21.

Smith, A.T., and Y. Xie. 2013. *Mammals of China*. Princeton, NJ: Princeton University Press.

Smith, W.C., and Butler, J.F. 1991. Ultrastructure of the Tabanidae compound eye: unusual features for diptera. *J Insect Physiol* 37:287–291, 293–296.

Speed, M.P., and G.D. Ruxton. 2007. How bright and how nasty: explaining diversity in warning signal strength. *Evolution* 61:623–635.

Spinage, C.A. 2012. *African ecology: benchmarks and historical perspectives*. New York: Springer.

Sponenberg, D.P. 2009. *Equine color genetics*. 3rd ed. Ames, IA: Wiley-Blackwell.

Stankowich, T. 2012. Armed and dangerous: predicting the presence and function of defensive weaponry in mammals. *Adapt Behav* 20:34–45.

Stankowich, T., and T. Caro. 2009. Evolution of weaponry in female bovids. *Proc R Soc Lond B Biol Sci* 276:4329–4334.

Stankowich, T., T. Caro, and M. Cox. 2011. Bold coloration and the evolution of aposematism in terrestrial carnivores. *Evolution* 65:3090–3099.

Stankowich, T., P. Havercamp, and T. Caro. 2014. Ecological drivers of antipredator defenses in carnivores. *Evolution* 68:1415–1425.

Steiner, C.C., and O.A. Ryder. 2011. Molecular phylogeny and evolution of the Perissodactyla. *Zool J Linn Soc* 163:1289–1303.

Stevens, M. 2013. *Sensory ecology, behavior, and evolution*. Oxford: Oxford University Press.

Stevens, M., I.C. Cuthill, A.M.M. Windsor, and H.J. Walker. 2006. Disruptive contrast in animal camouflage. *Proc R Soc Lond B Biol Sci* 273:2433–2438.

Stevens, M., and S. Merilaita. 2009a. Animal camouflage: current issues and new perspectives. *Philos Trans R Soc Lond B Biol Sci* 364:423–427.

———. 2009b. Defining disruptive coloration and distinguishing its functions. *Philos Trans R Soc Lond B Biol Sci* 364:481–488.

———. 2011. Animal camouflage: function and mechanisms. In *Animal camouflage: mechanisms and function*, eds. M. Stevens and S. Merilaita, 1–16. Cambridge: Cambridge University Press.

Stevens, M., and G.D. Ruxton. 2012. Linking the evolution and form of warning coloration in nature. *Proc R Soc Lond B Biol Sci* 279:417–426.

Stevens, M., W.T.L. Searle, J.E. Seymour, K.L.A. Marshall, and G.D. Ruxton. 2011. Motion dazzle and camouflage as distinct antipredator defenses. *BMC Biology* 9:81.

Stevens, M., D.H. Yule, and G.D. Ruxton. 2008. Dazzle coloration and prey movement. *Proc R Soc Lond B Biol Sci* 275:2639–2643.

Steverding, D., and T. Troscianko. 2004. On the role of blue shadows in the visual behavior of tsetse flies. *Proc R Soc Lond B Biol Sci* 271:S16–S17.

Stewart, D. 2004. *The zebra's stripes and other African animal tales.* Cape Town: Struik Publishers.

Stirton, R.A. 1940. Phylogeny of North American equidae. *Univ Calif Publ Geol Sci* 25:165–198.

St-Louis, A., and S.D. Cote. 2009. *Equus kiang* (Perissodactyla: Equidae). *Mamm Species* 835:1–11.

Stoner, C.J., T.M. Caro, and C.M. Graham. 2003. Ecological and behavioral correlates of coloration in artiodactyls: systematic attempts to verify conventional hypotheses. *Behav Ecol* 14:823–840.

Stuart, C., and T. Stuart. 1997. *Field guide to the larger mammals of Africa.* Cape Town: Struik Publishers.

Sundaresan, S.R., I.R. Fischhoff, J. Dushoff, and D.I. Rubenstein. 2007. Network metrics reveal differences in social organization between two fission-fusion species, Grevy's zebra and onager. *Oecologia* 151:140–149.

Sundaresan, S.R., I.R. Fischhoff, H.M. Hartung, P. Akilong, and D.I. Rubenstein. 2007. Habitat choice of Grevy's zebras (*Equus grevyi*) in Laikipia, Kenya. *Afr J Ecol* 46:359–364.

Sutcliffe, J.F. 1986. Black fly host location: a review. *Can J Zool* 64:1041–1053.

———. 1987. Distance orientation of biting flies to their hosts. *Insect Sci Applic* 8:611–616.

Tashiro, H., and H.H. Schwardt. 1949. Biology of the major species of horse flies of central New York. *J Econ Entomol* 42:269–272.

Tasker, M.L., P. Hope Jones, T. Dixon, and B.F. Blake. 1984. Counting seabirds at sea from ships: a review of methods employed and a suggestion for a standardized approach. *Auk* 101:567–577.

Tatin, L., S.R.B. King, B. Munkhtuya, A.J.M. Hewison, and C. Feh. 2009. Demography of a socially natural herd of Przewalski's horses: an example of a small, closed population. *J Zool* 277:134–140.

Taylor, D.B., R.D. Moon, and D.R. Mark. 2012. Economic impact of stable flies (Diptera: Muscidae) on dairy and beef cattle production. *J Med Entomol* 49:198–209.

Thaker, M., A.T. Vanak, C.R. Owen, M.B. Ogden, and R. Slotow. 2010. Group dynamics of zebra and wildebeest in a woodland savanna: effects of predation risk and habitat density. *PLOS ONE* 5:e12758.

Thayer, A.H. 1896. The law which underlies protective coloration. *Auk* 13:124–129.

———. 1918. Camouflage. *Scientific Monthly* 7:481–494.

Thayer, G.H. 1909. *Concealing-coloration in the animal kingdom: an exposition of the laws of disguise through color and pattern: being a summary of Abbott H. Thayer's discoveries*. New York: Macmillan.

Thompson, M.C. 1987. The effect on tsetse flies (*Glossina* spp.) of deltamethrin applied to cattle either as a spray or incorporated into ear-tags. *Trop Pest Mgmt* 33:329–335.

Thompson, P., and K. Mikellidou. 2001. Applying the Helmholtz illusion to fashion: horizontal stripes won't make you look fatter. *i-Perception* 2:69–76.

Thompson, V. 1973. Spittlebug polymorphic for warning coloration. *Nature* 242: 126–128.

Torr, S.J. 1989. The host-orientated behaviour of tsetse flies (*Glossina*): the interaction of visual and olfactory stimuli. *Physiol Entomol* 14:325–340.

———. 1994. Responses of tsetse flies (Diptera: Glossinidae) to warthog (*Phacochoerus aethiopicus* Pallas). *Bull Entomol Res* 84:411–419.

Torr, S.J., and J.W. Hargrove. 1998. Factors affecting the landing and feeding responses of the tsetse fly *Glossina pallidipes* to a stationary ox. *Med Vet Entomol* 12:196–207.

Tosh, C.R., J. Krause, and G.D. Ruxton. 2009. Basic features, conjunctive searches, and the confusion effect in predator-prey interactions. *Behav Ecol Sociobiol* 63:473–475.

Toupin, B., J. Huot, and M. Manseau. 1996. Effect of insect harassment on the behaviour of the Riviere George caribou. *Arctic* 49:375–382.

Trifonov, V.A., R. Stanyon, A.I. Nesterenko, B. Fu, P.L. Perelman, P.C.M. O'Brien, G. Stone, N.V. Rubtsova, M.L. Houck, T.J. Robinson, M.A. Ferguson-Smith, G. Dobigny, A.S. Graphodatsky, and F. Yang. 2008. Multidirectional cross-species painting illuminates the history of karyotypic evolution in Perissodactyla. *Chromosome Res* 16:89–107.

Trinkel, M. 2010. Prey selection and prey preferences of spotted hyenas *Crocuta crocuta* in the Etosha National Park, Namibia. *Ecol Res* 25:413–417.

Tullberg, B.S., S. Merilaita, and C. Wiklund. 2005. Aposematism and crypsis as a result of distance dependence: functional versatility of the colour pattern in the swallowtail butterfly larva. *Proc R Soc Lond B Biol Sci* 272:1315–1321.

Turing, A.M. 1952. The chemical basis of morphogenesis. *Philos Trans R Soc Lond B Biol Sci* 237:37–72.

Turnbull, W.D. 1970. Mammalian masticatory apparatus. *Fieldiana, Geology* 18:147–356.

Turner, D.A., and J.F. Invest. 1973. Laboratory analyses of vision in tsetse flies (Dipt., Glossinidae). *Bull Entomol Res* 62:343–357.

Tyler, S.J. 1972. The behaviour and social organization of the New Forest ponies. *Anim Behav Monogr* 5:85–96.

Uhlrich, D.J., E.A. Essock, and S. Lehmkuhle. 1981. Cross-species correspondence of spatial contrast sensitivity functions. *Behav Brain Res* 2:291–299.

Vale, G.A. 1974. The response of tsetse flies (Diptera: Glossinidae) to mobile and stationary baits. *Bull Entomol Res* 64:545–587.

Vale, G.A., and D.R. Hall. 1985a. The role of 1-octen-3-ol, acetone and carbon dioxide in the attraction of tsetse flies, *Glossina* spp. (Diptera: Glossinidae), to ox odour. *Bull Entomol Res* 75:209–217.

———. 1985b. The use of 1-octenol-3-ol, acetone and carbon dioxide to improve baits for tsetse flies, *Glossina* spp. (Diptera: Glossinidae). *Bull Entomol Res* 75:219–231.

Vale, G.A., D.F. Lovemore, S. Flint, and G.F. Cockbill. 1988. Odour-baited targets to control tsetse flies, *Glossina* spp. (Diptera: Glossinidae), in Zimbabwe. *Bull Entomol Res* 78:31–49.

Valeix, M., A.J. Loveridge, S. Chamaille-Jammes, Z. Devidson, F. Murindagomo, H. Fritz, and D.W. Macdonald. 2009. Behavioral adjustments of African herbivores to predation risk by lions: spatiotemporal variations influence habitat use. *Ecology* 90:23–30.

van Breugel, F., and M.H. Dickinson. 2012. The visual control of landing and obstacle avoidance in the fruit fly *Drosophila melanogaster*. *J Exp Biol* 215:1783–1798.

Vanderplank, F.L. 1942. A note on trypanosomiasis of game from tsetse areas at Shinyanga and Ukerewe peninsula. *Trans R Soc Trop Med Hyg* 35: 319–322.

———. 1944. Apparent densities of certain African blood-sucking insects (Diptera). *Proc Roy Ent Soc Lond A* 19:4–6.

Van Hennekeler, K., R.E. Jones, L.F. Skerratt, M.O. Muzari, and L.A. FitzPatrick. 2011. Meteorologcal effects on the daily activity patterns of tabanid biting flies in northern Queensland, Australia. *Med Vet Entomol* 25:17–24.

van Lawick, H. 1973. *Solo*. London: Collins.

van Lawick-Goodall, H., and J. van Lawick-Goodall. 1970. *Innocent killers: a fascinating journey through the worlds of the hyena, the jackal and the wild dog*. London: Collins.

van Orsdol, K.G. 1984. Foraging behavior and hunting success of lions in Queen Elizabeth National Park, Uganda. *Afr J Ecol* 22:79–99.

Vaughan, T.A. 1986. *Mammalogy*. Philadelphia: W.B. Saunders.

von Helversen, B., L.J. Schooler, and U. Czienskowski. 2013. Are stripes beneficial? dazzle camouflage influences perceived speed and hit rates. *PLOS ONE* 8:e61173.

Waage, J.K. 1981. How the zebra got its stripes: biting flies as selective agents in the evolution of zebra colouration. *J Ent Soc Sth Afr* 44:351–358.

Wakefield, S., J. Knowles, W. Zimmerman, and M. van Dierendonck. 2002. Status and action plan for the Przewalski's horse (*Equus ferus przewalskii*). In *Equids: zebras, asses and horses, status survey and conservation action plan*, ed. P.D. Moehlman, 82–92. Gland, Switzerland: IUCN.

Wallace, A.R. 1867a. Mimicry, and other protective resemblances among animals. *Westminster Foreign Q Rev* 31:1–43.

———. 1867b. *Proceedings of the Entomological Society of London*, March 4, IXXX–IXXXi.

———. 1877. The colours of animals and plants. Part I. *Am Nat* 11:384–408.

———. 1879. The protective colours of animals. *Science* 2:128–137.

———. 1891. *Natural selection and tropical nature*. London: Macmillan.

———. 1896. *Darwinism: an exposition of the theory of natural selection with some of its applications*. London: Macmillan.

Walsberg, G.E. 1983. Coat color and solar heat gain in animals. *BioScience* 33:88–91.

Walsberg, G.E., G.S. Campbell, and J.R. King. 1978. Animal coat color and radiative heat gain: a re-evaluation. *J Comp Physiol* 126:211–222.

Waltert, M., B. Meyer, M.W. Shanyangi, J.J. Balozi, O. Kitwara, S. Qolli, H. Krishke, and M. Muhlenberg. 2008. Foot surveys of large mammals in woodlands of western Tanzania. *J Wildl Mgmt* 72:603–610.

Warmuth, V., A. Eriksson, M.A. Bower, G. Barker, E. Barrett, B.K. Hanks, S. Li, D. Lomitashvilli, M. Ochir-Goryaeva, G.V. Sizonov, V. Soyonov, and A. Manica. 2012. Reconstructing the origin and spread of horse domestication in the Eurasian steppe. *Proc Nat Acad Sci* 109:8202–8206.

Washino, R.K., and C.H. Tempelis. 1983. Mosquito host bloodmeal identification: methodology and data analysis. *Ann Rev Entomol* 28:179–201.

Webb, S.D. 1977. A history of savanna vertebrates in the New World. Part I, North America. *Ann Rev Ecol Syst* 8:355–380.

Webster R.J., C. Hassall, C.M. Herdman, J-G. J. Godin, and T.N. Sherratt. 2013. Disruptive camouflage impairs object recognition. *Biol Lett* 9: 20130501.

Weinstock, J., E. Willerslev, A. Sher, W. Tong, S.Y.W. Ho, D. Rubenstein, J. Storer, J. Burns, L. Martin, C. Bravi, A. Prieto, D. Froese, E. Scott, L. Xulong, and A. Cooper. 2005. Evolution, systematics, and phylogeography of Pleistocene horses in the New World: a molecular perspective. *PLOS Biology* 3:e241.

Weitz, B. 1963. The feeding habits of *Glossina*. *Bull World Health Organ* 28:711–729.

Welburn, S.C., and I. Maudlin. 1991. Rickettsia-like organisms, puparial temperature and susceptibility to trypanosome infection in *Glossina morsitans*. *Parasitology* 102:201–206.

———. 1999. Tsetse-trypanosome interactions: rites of passage. *Parasitology Today* 15:399–403.

Wells, S.M., and B. van Goldschmidt-Rothschild. 1979. Social behaviour and relationships in a herd of Camargue horses. *Z Tierpsychol* 49:363–380.

West, P.M., and C. Packer. 2002. Sexual selection, temperature, and the lion's mane. *Science* 297:1339–1343.

West, S. 2009. *Sex allocation*. Princeton, NJ: Princeton University Press.

Wichman, H.A., C.T. Payne, O.A. Ryder, M.J. Hamilton, M. Maltbie, and R.J. Baker. 1991. Genomic distribution of heterochromatic sequences in equids: implications to rapid chromosomal evolution. *J Hered* 82:369–377.

Wiesenhutter, E. 1975. Research into the relative importance of Tabanidae (Diptera) in mechanical disease transmission. II. Investigation of the behavior and feeding habits of Tabanidae in relation to cattle. *J Nat Hist* 9:385–392.

Willett, K.C. 1960–1961. Recent developments in research on the African trypano-somiases. *Trans New York Acad Sci* 23:233–236.

Williams, D. 2001. *Naval camouflage 1914-1945: a complete visual reference*. Annapolis, MD: Naval Institute Press.

Williams, S.D. 2002. Status and action plan for Grevy's zebra (*Equus grevyi*). In *Equids: zebras, asses and horses, status survey and conservation action plan*, ed. P.D. Moehlman, 11–27. Gland, Switzerland: IUCN.

———. 2013. *Equus grevyi*, Grevy's zebra. In *Mammals of Africa*, vol. 5, *Carnivores, pangolins, equids and rhinoceroses*, eds. J. Kingdon and M. Hoffmann, 422–428. London: Bloomsbury.

Wilson, D.E., and D.M. Reeder. 1993. *Mammal species of the world: a taxonomic and geographic reference*. 3rd ed. Baltimore: Johns Hopkins University Press.

Wohlfender, F. 2009. Equine encephalosis—an emerging threat. *DEFRA/AHT/ BEVA Equine Quarterly Disease Surveillance Report* 5:8–12.

Wolski, T.R., K.A. Houpt, and R. Anderson. 1980. The role of the senses in mare-foal recognition. *Appl Anim Ethol* 6:121–138.

World Organization for Animal Health (OIE). 2009. Trypanosomosis (tsetse-transmitted). OIE Technical Disease Cards. Accessed 25 April 2013. http:// www.oie.int/fileadmin/Home/eng/Animal_Health_in_the_World/docs/pdf /VEE_FINAL.pdf.

Wunderer, H., and U. Smola. 1982. Fine structure of ommatidia at the dorsal eye margin of *Calliphora erythrocephala* Meigen (Diptera: Calliphoridae): an eye region specialized for the detection of polarized light. *Int J Insect Morphol Embryol* 11:25–38.

Xu, F., M. Ma, W. Yang, D. Blank, P. Ding, and T. Zhang. 2013. Group size effect on vigilance and daytime activity budgets of the *Equus kiang* (Equidae, Perissodactyla) in Arjinshan National Nature Reserve, Xinjiang, China. *Folia Zool* 62:76–81.

Yamaguchi, Y., M. Brenner, and V.J. Hearing. 2007. The regulation of skin pigmentation. *J Biol Chem* 282:27557–27561.

Yapic, O., S. Yavru, M. Kale, O. Bulut, A. Simsek, and K.C. Sahna. 2007. An investigation of equine infectious anemia infection in the Central Anatolia region of Turkey. *S Afr Ve Ver* 78:12–14.

Zeise, L., M.R. Chedekel, and T.B. Fitzpatrick. 1995. *Melanin: its role in human photoprotection*. Overland Park, KS: Valdenmar.

Zhirnov, L., and V. Ilyinsky. 1986. *The Great Gobi National Park: refuge for rare animals of the central Asian deserts*. Moscow: Centre for International Projects.

Zollner, G.E., S.J. Torr, C. Ammann, and F.X. Meixner. 2004. Dispersion of carbon dioxide plumes in African woodland: implications for host-finding by tsetse flies. *Physiol Entomol* 29:381–394.

Zumpt, F. 1973. *The Stomoxyine biting flies of the world. Diptera: Muscidae. Taxonomy, biology, economic importance and control measures*. Stuttgart, Germany: Gustav Fischer Verlag.

Index